Spark Ignition Engine Modeling and Control System Design

This book presents a step-by-step guide to the engine control system design, providing case studies and a thorough analysis of the modeling process using machine learning, and model predictive control (MPC). Covering advanced processes alongside the theoretical foundation, MPC enables engineers to improve performance in both hybrid and non-hybrid vehicles.

Control system improvement is one of the major priorities for engineers seeking to enhance an engine. Often possible on a low budget, substantial improvements can be made by applying cutting-edge methods, such as artificial intelligence when modeling engine control system designs and using MPC. This book presents approaches to control system improvement at mid, low, and high levels of control. Beginning with the model-in-the-loop hierarchical control design of ported fuel injection SI engines, this book focuses on optimal control of both transient and steady state and also discusses hardware-in-the-loop. The chapter on low-level control discusses adaptive MPC and adaptive variable functioning, as well as designing a fuel injection feed-forward controller. At mid-level control, engine calibration maps are discussed, with consideration of constraints such as limits on pollutant emissions. Finally, the high-level control methodology is discussed in detail in relation to transient torque control of SI engines.

This comprehensive yet clear guide to control system improvement is an essential read for any engineer working in automotive engineering and engine control system design.

Spark Ignition Engine Modeling and Control System Design

A Guide to Model-in-the-Loop Hierarchical Control Methodology

Amir-Mohammad Shamekhi

Amir Hossein Shamekhi

CRC Press is an imprint of the
Taylor & Francis Group, an **informa** business

First edition published 2023
by CRC Press
6000 Broken Sound Parkway NW, Suite 300, Boca Raton, FL 33487-2742

and by CRC Press
4 Park Square, Milton Park, Abingdon, Oxon, OX14 4RN

CRC Press is an imprint of Taylor & Francis Group, LLC

ISBN: 978-1-032-34612-0 (hbk)
ISBN: 978-1-032-34615-1 (pbk)
ISBN: 978-1-003-32304-4 (ebk)

DOI: 10.1201/9781003323044

Typeset in Times
by KnowledgeWorks Global Ltd.

Dedication

To the greatest treasures of my life;

To my beloved mother,
from whom I learnt affection, courtesy, and benevolence;
and to my dearest father,
who taught me devotion, compassion, and diligence,
who is watching over me,
somewhere up there, from Heaven.

—Amir-Mohammad Shamekhi

To:

My late father,
my mother,
my wife, Saeedeh,
my son, Mohammad-Mehdi,
and my daughter, Ghazal.

—Amir Hossein Shamekhi

Contents

List of Abbreviations

NOMENCLATURE

PARAMETERS

a	Ratio of plate rod diameter to the plate diameter
A	Effective intake area of throttle body
A_D	Frontal area
a_x	Vehicle acceleration
C_d	Discharge coefficient; Drag coefficient
c_P	Specific heat at constant pressure
c_{us}	Turbine blade speed ratio (in Turbocharger)
$\tilde{c}_{us}$	Turbine corrected blade speed ratio (in Turbocharger)
c_v	Specific heat at constant volume
D	Inside diameter of throttle
D_A	Aerodynamic drag force
d_c	Diameter of compressor blade
$E(w(k))$	Network error function in the kth iteration
$g(n)$	Gradient vector of the error function
H	Enthalpy
$H(n)$	Hessian matrix
J	Objective function
$J(n)$	Jacobian matrix
K	Specific heat ratio of air
L	Load
M	Engine torque vector; vehicle mass
M_a	Mach number
M_{eq}	Equivalent mass
$\dot{m}$	Mass rate
$\dot{m}_{air}$	Air mass flow rate
$\dot{m}_{in}$	Mass flow rates into manifold
$\dot{m}_{out}$	Mass flow rates out of manifold
$m_{x,lim}$	Standard limit for pollutant x
$\overline{m}_{x,NEDC}$	The sum of pollutant x over the course of NEDC
$\widehat{\dot{m}}_a$	Estimated air mass flow rate into the cylinder
$\dot{m}_a$	Actual air mass flow rate into the cylinder
$\dot{m}_{air}$	Air mass flow rate
$\dot{m}_{fuel}$	Fuel mass flow rate
$m_f(t)$	Mass of fuel film
$\dot{m}_{inj}$	Injected fuel mass flow rate
M_r	Equivalent mass of the rotating components
$\dot{m}_\psi(t)$	Injected fuel mass
$\dot{m}_\varphi(t)$	Fuel mass flow rate into the cylinder

n	Engine speed, in rpm
N	Engine speed vector
n_{cyl}	Number of cylinders
n_i	Engine speed, at steady-state point i
N_{tf}	Ratio of the combined transmission and final drive
P_c	Compressor power (in turbocharger)
P_{cr}	Critical pressure
P_{in}	Ambient pressure
P, P_m, P_{out}	Intake manifold pressure
P_t	Turbine power (in turbocharger)
P_0	Ambient pressure
Q	Heat transfer
r	Rolling radius of the tires
R	Gas constant
R_{hx}	Hitch force
R_x	Rolling resistance force
T	Engine torque
t_{inj}	Injection duration
T_c	Compressor torque (in turbocharger)
T_e	Generated brake torque
T_f	Friction loss torque (in turbocharger)
T_i	Engine Torque, at steady-state point i
T_l	External load
T_t	Turbine torque (in turbocharger)
u	Control command; control input to be optimized
$u(n)$	System input vector at discrete time n
U	Control command vector
U_c	Corrected speed, (in Turbocharger)
V	Manifold volume
V_d	Displacement volume
W	Vehicle weight
w_x	Weighting factor for constituent x
$w(n)$	Network weight matrix in the nth iteration
$x(n)$	State vector at discrete time n
y	Estimated air mass flow rate into the cylinder
$y(n)$	System output vector at discrete time n
y'	Actual fuel mass flow rate into the cylinder
$\hat{y}$	Neural network output
y_u	Real system output
α	Convective heat transfer coefficient (in Turbocharger)
α_e	Engine rotational acceleration
Δ	Variation, difference
δ_t	Time increment
ε	Multiplicative unstructured uncertainty
η	Learning rate

η_c	Compressor isentropic efficiency (in turbocharger)
c	Engine efficiency
η_t	Turbine isentropic efficiency (in turbocharger)
η_{tf}	Combined efficiency of the transmission and final drive
ϑ	System temperature, manifold instantaneous temperature
ϑ_0	Ambient temperature
θ_0	Angle of throttle plate when closed
θ_i	Engine throttle angle, at steady-state point i
Θ_e	Rotational inertia
ϑ_{in}	Ambient or upstream temperature
Θ_{tc}	Turbocharger rotational inertia
θ_{thr}	Throttle angle
κ	Magnifier
λ	Lambda, or the ratio of actual AFR to stoichiometric AFR
λ_l	Volumetric coefficient
$\dot{\mu}_c$	Compressor corrected mass flow rate (in turbocharger)
$\dot{\mu}_t$	Turbine corrected mass flow rate (in turbocharger)
Π_c	Compressor inlet–outlet pressure ratio (in turbocharger)
Π_t	Turbine inlet-outlet pressure ratio (in turbocharger)
ρ_{in}	Density of the fluid into the cylinder
σ	Stoichiometric fuel-air ratio
τ	Time constant
τ_{IEG}	Induction-to-exhaust-gas delay
τ_{IPS}	Induction-to-power-stroke delay
τ_{seg}	Segment time, or the ECU sampling time
φ	Air-fuel equivalence ratio
Φ	Normalized compressor flow rate
φ_{ign}	Ignition angle
$\tilde{\omega}_c$	Compressor corrected speed (in turbocharger)
$\omega_e(t)$	Engine speed, rad/s
$\tilde{\omega}_t$	Turbine corrected speed (in turbocharger)
ω_{tc}	Turbocharger shaft speed
Ω_n	Speed weighting factor matrix
Ω_T	Torque weighting factor matrix
Ω_u	Control command weighting factor matrix

Abbreviations

ABC	Artificial Bee Colony
AC	Ant colony
AFR	Air-fuel ratio
ANFIS	Adaptive neuro-fuzzy inference system
ATDC	After top dead centre
AVF	Adaptive variable functioning
AVR	Adaptive variable rate

BSFC	Brake-specific fuel consumption
BTDC	Before top dead centre
CA	Crank angle
CFD	Computational fluid dynamics
CK	Cuckoo-search method
CNN	Convolutional neural network
CO	Carbon monoxide
CRDI	Common rail direct injection
DC	Driving cycle
DE	Differential Evolution
DEM	Discrete event modeling
DI Diesel	Direct injection diesel
DRNN	Diagonal recurrent neural networks
DTM	Discrete time modeling
ECU	Electronic control unit
EGR	Exhaust gas recirculation
EMS	Energy management system
FLOPS	Floating-point operations per second
FTP	Federal test procedure
GA	Genetic algorithm
HC	(Unburnt) Hydro-Carbon
HCCI	Homogeneous charge compression ignition
HiL	Hardware-in-the-loop
IC Engine	Internal combustion engine
IMEP	Indicated mean effective pressure
LOLIMOT	Local linear model tree
LOPOMOT	Local polynomial model tree
LQ	Linear quadratic
MAP	Manifold absolute pressure
MiL	Model-in-the-loop
MIMO	Multi-input multi-output
MLP	Multi-layer perceptron
MP	Model predictive
MPC	Model predictive control
MSE	Mean squared error
MVM	Mean value modeling
NA	Naturally aspirated
NARX	Nonlinear autoregressive model with exogenous input
NEDC	New European Driving Cycle
NMPC	Nonlinear model predictive control
NN	Neural network
NOx	Nitrogen oxides
PCA	Principal component analysis
PI	Proportional-integral (controller)
PIP	Proportional-integral-plus (controller)

PRBS	Pseudo-random binary signal
PSO	Particle swarm optimization
RBF	Radial basis function
RCP	Rapid control prototyping
RHP	Right-hand-plane
RLS	Recursive least square
RNN	Recurrent neural network
rpm	Revolution per minute
SA	Spark advance
SI	Spark ignition (engine)
SiL	Software-in-the-loop
SISO	Single-input single-output
STR	Self-tuning regulator
TDC	Top dead center
TPU	Time processing unit
TW	Three-way
TWC	Three-way catalytic converter
UHC	Unburned hydrocarbons
VCT	Variable camshaft timing
VGT	Variable geometry turbocharger
VVT	Variable valve timing
WLTP	World harmonized light vehicle test procedure

Preface

The automobile is an indispensable part of human life. Meanwhile, improvement and optimization of internal combustion (IC) engines, as the primary source of propulsion in the majority of vehicles, plays a significant role in the competitive arena of the automotive industry. With emission standards becoming more and more stringent, this enhancement is strongly pursued in three major realms of fuel consumption reduction, pollutant emission abatement, and performance improvement.

According to studies, by 2050, a notable proportion of the vehicles would be either hybrid or conventional.[i] Consequently, IC engines would still remain the main power generation source for a large number of automobiles.

Of all the engine enhancement procedures, control system improvement is of major priority, owing to lower cost and substantial impact on the aforementioned objectives.

This book is to present a simple step-by-step guide for automotive engineers, students, and researchers as to engine control system design, from modeling to controller design, with a new perspective, through a case study. A new approach in the model-in-the-loop (MiL) hierarchical control design of ported fuel injection SI engines is introduced, with focus on optimal control of both transient and steady-state behaviors. As will be explained, transients play major role in newer driving cycles and hybrid vehicles and are usually not given appropriate importance in the literature.

MiL is the first step in control design and the pre-requisite for other steps like hardware-in-the-loop (HiL). MiL consists of a control-oriented model and the hierarchical control structure, which per se, comprises high-level, mid-level, and low-level layers.

After the introduction and literature survey in Chapter 1, control-oriented modeling is discussed in Chapter 2. Using the gray-box structure, an engine global model/simulator with high accuracy and fast response is presented. Applying a part-to-whole approach (inspired from mean value modeling [MVM]), the SI engine system is decomposed to simpler subsystems, and a combination of white-box and black-box modeling is utilized to bring both accuracy and reliability. An effective handy instruction for design of experiments is presented, and an adequate data acquisition is performed. Using effective techniques, such as committee structures, simplifying the tasks burdened on neural networks, and local-global modeling, highly accurate identifiers are achieved. The final complete model, named Neuro-MVM, successfully accounts for engine dynamic behavior, in a variety of ambient circumstances, for manifold pressure, engine speed, fuel consumption, produced emissions, knock detection, and after-catalyst emissions, with high accuracy in wide ranges of operation for inputs and outputs, which are notable achievements. The methodology presented can be extended to other types of engines.

With regard to mid-level control, Chapter 3 presents a model-based calibration for reducing fuel consumption and pollutant emissions with a new perspective. Instead of conventional local modeling, the global model designed is employed for optimization. As will be explained, this relieves the need for harmful harsh smoothening required for some local models. A driving-cycle-based calibration is performed with after-catalyst emissions and knock limits taken into consideration. Both part-load and full-load modes of engine in the widest ranges of engine operation will be covered. To eliminate variance errors resulted from meta-heuristic optimization, the committee method will be applied, which unlike many other approaches does not distort the original results. In the end, mid-level controller calibration maps for spark advance, throttle angle, air-fuel ratio (for full-load mode), and VVT are presented in widest ranges of engine operation.

As regards fuel injection feed-back (low-level) control, a model predictive self-tuning regulator (MP-STR) with adaptive variable functioning (AVF) is designed in Chapter 4. Proper control of wall-wetting effects in transients is a matter of great challenge in ported fuel injection SI engines. Self-tuning regulator is employed to address both parametric and unstructured uncertainties, but in a predictive practice, in order to achieve optimal behavior, particularly in transient phases. Additionally, it is applied with AVF; namely, at low speeds with longer computation time available, the MP-STR is wholly applied, yet at higher speeds with shorter computation time in hand, a simplified lighter controller is employed. As will be seen, the control system is fully capable of maintaining the output close to the desired value, much more preferable than that of a conventional controller.

As far as high-level control is concerned, a new methodology regarding transient torque control of SI engines is presented in Chapter 5. Transients are of great importance especially for hybrid vehicles as the engine is frequently started and loaded up to demanded torques. To address bandwidth requirements, purely feedback controllers are avoided, and a semi-feed-forward controller is incorporated into high-level layer of the hierarchical structure. The controller is derived based on off-line nonlinear model predictive control, minimizing the fuel consumption for optimal transition to demanded torques.

It shall be reiterated that the focus of this book is on model-in-the-loop control design, which is the basis for subsequent design steps, e.g. HiL design. More information regarding these next steps could be found in the references discussed in this book.

We warmly welcome and appreciate remarks, criticisms, and suggestions for future editions from all scholars, engineers, researchers, and students in the field.

NOTE

i. For instance, refer to EESI. *Transportation 2050: More EVs, but Conventional Vehicles Will Still Dominate.* Available from: https://www.eesi.org/articles/view/transportation-2050-more-evs-but-conventional-vehicles-will-still-dominate.

Author

Amir-Mohammad Shamekhi received his PhD in Automobile Control from K. N. Toosi University of Technology, Tehran, Iran, in 2021, and was chosen as the superior researcher of the Department of Mechanical Engineering. His works concern Control and Machine Learning, and their applications particularly in vehicles, about which he has published in several journal papers.

Amir Hossein Shamekhi received his B.Sc. in Mechanical Engineering from Tehran University, Tehran, Iran, in 1993. Carrying on his studies, he obtained M.Sc. from K. N. Toosi University of Technology, Tehran, Iran, in 1997. Receiving his PhD in 2004, Dr. Shamekhi was the first PhD alumni of Mechanical Engineering in K. N. Toosi University of Technology, where he is currently Associate Professor in the faculty of Mechanical Engineering. His fields of study include internal combustion engines, mechatronics, and automotive transmission.

1 Introduction

1.1 ENGINE CONTROL SYSTEM DEVELOPMENTS

As explained in the Preface, engine control system improvement will still play an important role for a portion of automobiles in the next two decades or so. Until 1965, internal combustion (IC) engines were controlled mechanically via transistor-triggered electromechanical coil ignition. Gradually, with the advent of fuel injectors (1967), three-way catalysts (1973), and oxygen sensors (1976), control systems started to be increasingly introduced for combustion engines. These developments demanded more sensors (e.g. knock sensor, pressure sensor, etc.) with electrical outputs, actuators with electrical inputs (e.g. fuel injectors), and thus, a microprocessor (1979). In particular, the introduction of emission regulations (starting from EURO#1, 1992) led to more and more improvements in engine control systems. Future developments will undoubtedly invoke more complicated control systems [1].

Commonly, the electronic control unit (ECU) of IC engines includes a microprocessor and perhaps a time processing unit (TPU), whose task is to coordinate control commands with the reciprocal behavior of the engine (Figure 1.1). In terms of memory and processing speed, ECUs are much weaker than personal computers, as they need to be of greater durability and robustness to work in harsh and variable environments [2].

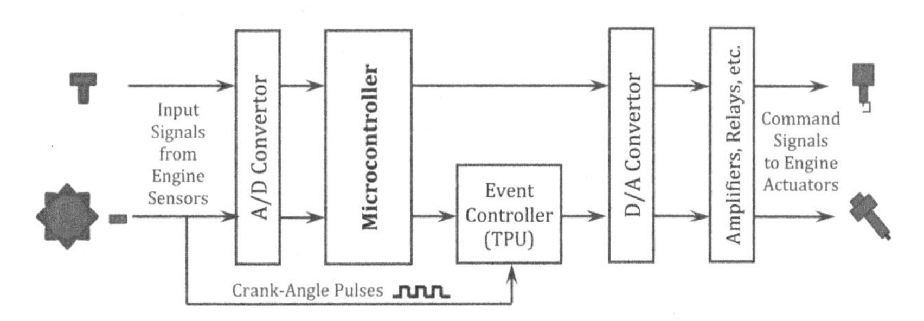

FIGURE 1.1 Engine control unit structure [2].

1.2 DESIGN AND SIMULATION OF CONTROL SYSTEMS

Control system design and simulation can be carried out in four steps [1, 3]:

- **Model-in-the-loop (MiL):** In the first development phase, control design could be based on simulations including an engine model and

DOI: 10.1201/9781003323044-1

an ECU model, namely, control algorithms in a high-level language, e.g. MATLAB-Simulink. This is basically the first step in control design.

- **Software-in-the-loop (SiL):** Quite similar to MiL, SiL includes a simulated process with a simulated controller. A SiL simulation compiles the generated code and executes it as a separate process on the host computer. During SiL simulation, code coverage and execution-time metrics of the generated code could be collected [4]. SiL is the preliminary step for HiL.
- **Hardware-in-the-loop (HiL):** It contains a real-time engine model, with real ECU and real components.
- **Control prototyping:** It contains a real engine with a real ECU. If some new control functions have been developed, they can be applied to an available ECU in the *bypass mode*. The new control functions could be tested with a special real-time computer in parallel to the ECU. This is called *rapid control prototyping* (RCP).

Figure 1.2 displays the above different structures.

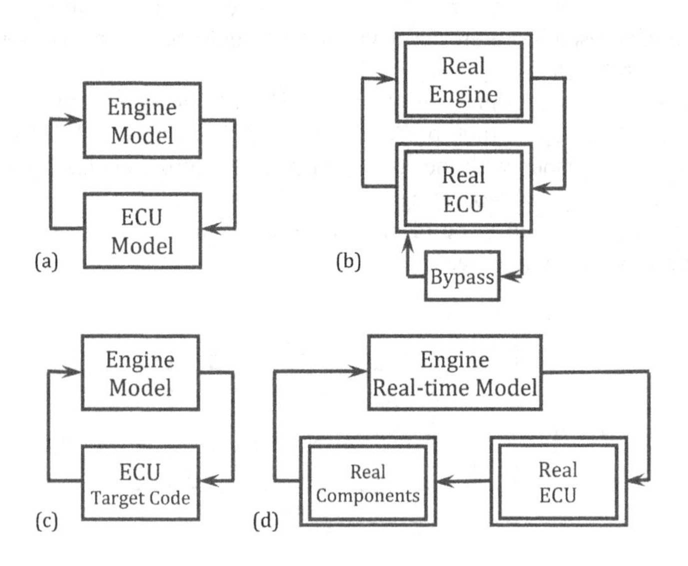

FIGURE 1.2 Different simulation and design architectures for control functions and software development: (a) MiL simulation, (b) RCP, (c) SiL simulation, and (d) HiL simulation [1].

Design and simulation of the control system in SiL and HiL clearly leads to a substantial reduction in trial and error, and thus the amount of cost and time required, and also results in an improvement of the final product. It is to be noted that actuators are not necessarily modeled in MiL and SiL, and would be addressed in later steps, e.g. HiL. More information thereof can be found in [3, 5].

This study is to present a MiL control design for a spark ignition (SI) engine. Accordingly, as discussed above, this demands an engine control-oriented model with a control structure. Figure 1.3 portrays the schematic of a port fuel injection SI engine and its relations with the ECU. Figure 1.4 depicts a simplified block diagram of the engine control structure.

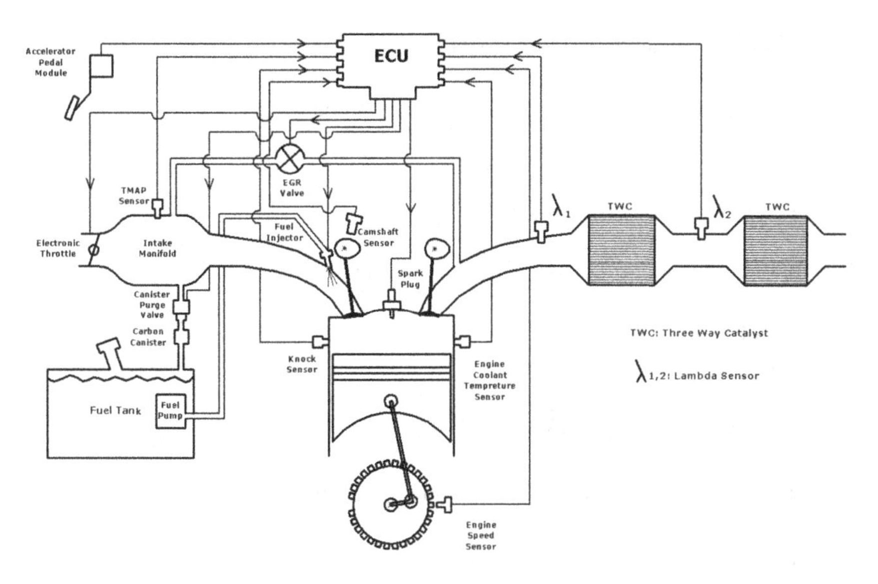

FIGURE 1.3 The schematic of a port fuel injection SI engine and its relations with the ECU.

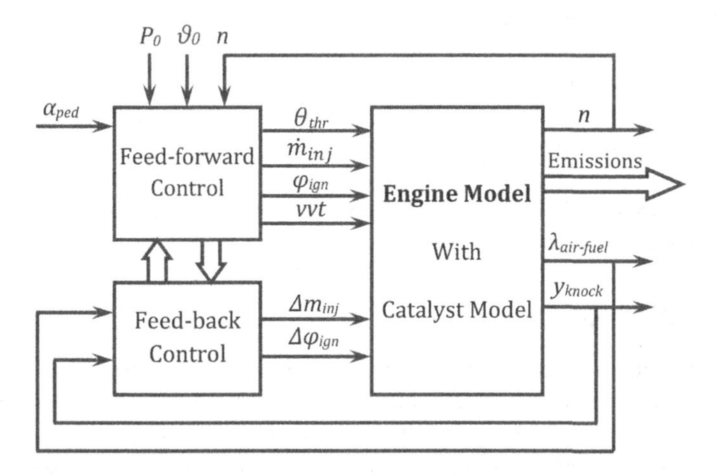

FIGURE 1.4 A simplified schematic of control structure in a port fuel injection SI engine (inspired from [1]).

As can be seen, the engine controller is a combination of feed-forward and feed-back controllers. The inputs to the feed-forward controller box include:

- Ambient pressure (P_0)
- Ambient temperature (ϑ_0)
- Engine speed (n, fed back from an rpm sensor)
- Driver's command (α_{Ped}, gas pedal angle)
- Manifold absolute pressure (MAP) sensor

The inputs to the feed-back controller block include:

- Lambda sensor signal ($\lambda_{\text{air-fuel}}$)
- Knock sensor signal (y_{knock})

The engine inputs coming from feed-forward controllers include:

- Throttle angle (θ_{thr})
- Fuel mass injected (m_{inj})
- Spark advance (SA) (φ_{ign})
- Variable valve timing (VVT)

The engine inputs from feed-back controllers include:

- Fuel injection correction (Δm_{inj})
- SA ($\Delta\varphi_{\text{ign}}$)

Engine outputs include:

- rpm sensor signal (n)
- Pollutant emissions
- MAP sensor signal
- Lambda sensor signal ($\lambda_{\text{air-fuel}}$)
- Knock sensor signal (y_{knock})

In what follows, a short glance will be given at the two major MiL parts, i.e. engine modeling and control.

1.3 MODELING

Modeling is the first step in the majority of control designs, especially for MiL development. A simple schematic of the inputs and outputs of an SI engine is depicted in Figure 1.5. As for engines, the model shall indicate the effects of engine inputs on engine outputs, both *accurately* and in *real time*, and could be employed in HiL, SiL, or MiL developments of engine ECU and/or energy management system (EMS) of a hybrid vehicle. Any flaw in the aforementioned two features (i.e. accuracy and fast response) would degrade the performance of the controller (ECU and EMS) to be designed. On top of these, a model (on a higher level) in the role of a simulator can be utilized for the system investigation and improvement and can significantly reduce the amount of required experiment, time, and cost. Therefore, designing engine models with more accuracy can result in the optimal use of fuel, reduction of pollutant emissions, and better performance at a lower cost.

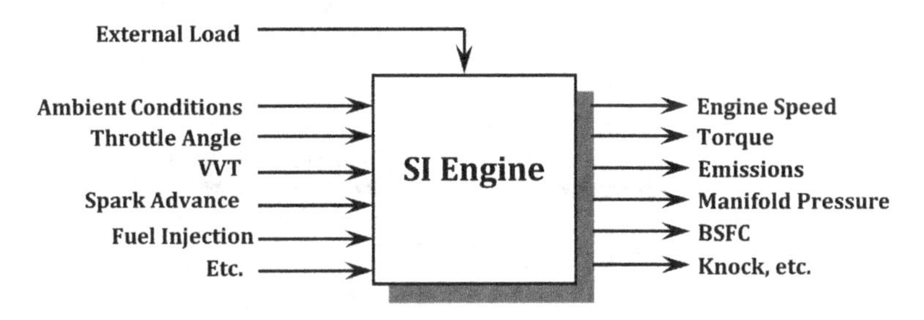

FIGURE 1.5 SI engine inputs and outputs.

An internal combustion engine is a highly nonlinear and complicated system. Modeling based on solving physical equations running in an engine (i.e. thermo-fluid modeling), while accurate, is too time-consuming and thus not appropriate for a control purpose. As a matter of fact, designing a control system does involve a *control-oriented model*, which is real time. *Mean value modeling* (MVM) is the most commonly used method of engine modeling and underpins many other control-oriented modeling approaches [2, 3].

MVM attempts to simplify the highly complicated physical phenomena and mathematical equation sets governing the real engines, in order to reduce the number of calculations required and achieve a fast-response model. In MVM, the reciprocal behavior of an engine is ignored, and it is assumed like a turbine continuously producing torque [2]. The engine is decomposed into its subsystems, such that the simplest individual tasks are achieved. Depending on their time constant, the subsystems are categorized as static (with rapid response), dynamic (with medium response), and constant (with too slow response). They can be described, respectively, with algebraic equations, first-order ordinary differential equations, or constant values. The principal engine subsystems in MVM accompanied with their status are given as follows [2]:

- Throttle body (static)
- Gas exchange (static)
- Combustion (static)
- Catalytic converter (dynamic)
- Engine temperature (semi-constant)
- Intake manifold (dynamic)
- Wall-wetting (dynamic)
- Exhaust manifold (dynamic)
- Engine inertia (dynamic)
- Turbocharger (if applicable)

Figure 1.6 represents the SI engine subsystems with their inputs and outputs in MVM. As can be seen, the main engine inputs of an SI engine include throttle angle, ambient air conditions, injected fuel, SA, VVT, and external load (or the disturbance).

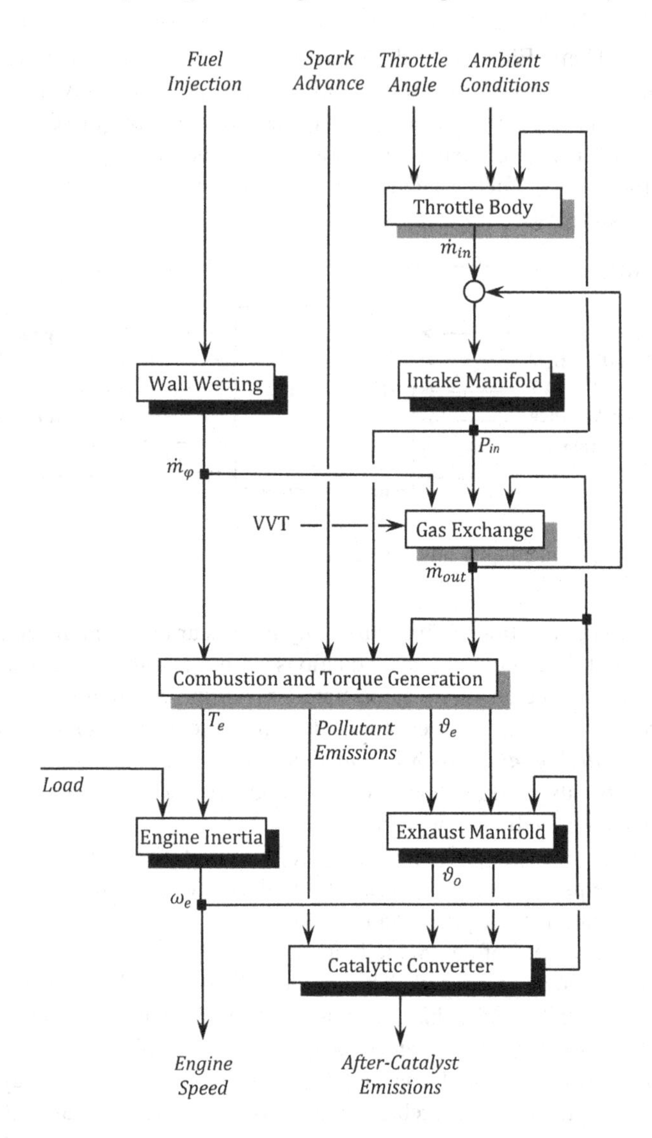

FIGURE 1.6 SI engine subsystems with their inputs and outputs in MVM (inspired from [2]). Note that the turbocharger and engine temperature subsystems are not shown. Besides, the exhaust manifold subsystem could have different input/output arrangements. The blocks with black shadings are dynamic, while the blocks with gray shadings are static.

Note that MVM does suffice for a MiL design. Yet, when it comes to a HiL step (which is beyond the scope of this book), the reciprocal behavior of the engine needs to be taken into consideration, leading to discrete event modeling (DEM).

In DEM, instead of a (continuous) time domain, the system is represented in a crank-angle domain. In order to achieve linear equations, a crank-angle domain is desired to be transformed into a discrete-time domain. Note that the crank-angle duration equals engine speed multiplied by time duration. Furthermore, as a matter of fact, the intake manifold dynamic is of the order of at least 10 times faster than that of the engine inertia. Consequently, engine speed could be assumed constant in small intervals of engine operation. Accordingly, in small intervals (e.g. 10 cycles), sampling time could be assumed constant, allowing the system to be solved in a (discrete) time domain rather than a crank-angle domain. In other words, although the model is defined in the crank-angle domain, it is formulated on the basis of constant sampling times. This, per se, leads DEM to be converted to discrete-time modeling (DTM), significantly simplifying the formulations and linearizing many of them.

DTM consists of various continuous-time and discrete-time subsystems. The basic sampling time for DTM is defined as the time duration between two consecutive plain (i.e. with zero advance or retard) ignition events and is called segment time (). More information regarding discrete modeling can be found in [2].

In MVM, the discrete reciprocal behavior of the engine is captured by introducing (at least) two major time delays: the induction-to-power-stroke delay and the induction-to-exhaust-gas delay. These will be explored in the respective section (see Section 2.4.5).

On the other hand, mean value models, from the very beginning [6, 7] until now [1, 2], have been suffering from several major drawbacks. As mentioned, these models were proposed to speed up the time response of engine models. However, this was realized at the price of almost losing the model's accuracy. In addition, these models are weak at predicting pollutant emissions. Some others combined mean value models and thermodynamic models [8]. While this action substantially improved the model accuracy, the model speed significantly declined. In fact, the defects of mean value models lie behind the simplification of intrinsic complexities of the engine's physics, in order to obtain a fast-response model. That is to say, the problem is the numerous unstructured uncertainties in MVM.

Consequently, the idea of utilizing black-box models was proposed. Black-box models are designed merely based on empirical input/output data and are not involved in the complications of solving highly nonlinear sets of equations.[i] Classical black-box techniques typically required an abundant number of equidistant empirical data. This necessity significantly reduced after the advent of artificial neural networks [9], as they are fairly intelligent and demand lesser experimental data.

Most of the papers reviewed, concerning engine modeling with neural networks, only concentrate on some features of combustion (such as heat release or pollutants' prediction), not modeling the entire engine. Muller and Hemberger [10] employed a static neural network to predict the point of 50% heat release in real time. An adaptive ignition control system was erected based on this model. Gang et al. [11] employed one dynamic neural network to model a diesel engine. The model predicted the engine speed in a range of 1300 rpm. He and Rutland modeled some of the steady-state characteristics of combustion using neural networks. They utilized notable methods in designing neural networks [12, 13]. Nonetheless, their methodology for generalizing the steady-state results of combustion to the dynamic behavior of the whole engine was defective [14]. Brahma et al. used static neural networks to predict cylinder pressure [15]. Their methods for enhancing the performance of neural networks were again remarkable.

Brusca et al. employed a neural network to estimate the variable-specific heat ratio for a single-zone model. The model was to indicate the heat release in a cylinder [16]. Samadani (2009) established a combustion model for a multi-objective optimization purpose using a static neural network. The results were used to pre-calibrate the engine emissions and performance [17]. Shivakumar and Shrinivasa Rao attempted to indicate the steady-state characteristics of a variable compression ratio, diesel engine using two static neural networks [18]. Uzun employed a neural network as a tool to proliferate the results attained from an engine test and to conduct a parametric study [19]. Yap and Karri examined three feed-forward networks to predict the steady-state emissions of an engine. They found optimization layer-by-layer with better results than back-propagation, and radial basis function networks [20]. Ismail et al. suggested a static neural network model for a diesel engine. They utilized 320 patterns to train a neural network with four inputs and seven outputs [21]. The results could have become better with a larger pattern table and distributing the seven outputs between seven single-output networks (as shown in [13]). This problem is resolved in Cay et al.'s work. They presented a static, neural network model for a methanol engine [22]. Janakiraman et al., in a notable work, tried to model some of the transient characteristics of a homogeneous charge compression ignition (HCCI) engine combustion process. They employed the series-parallel network to predict one step ahead and the nonlinear autoregressive exogenous (NARX) network to estimate multi-steps ahead. They found the multi-layer-perceptron (MLP) structure more favorable than the radial-basis network (RBN). They also decreased the input dimension (and thus the memory required) using principal component analysis (PCA) [23].

Molina et al. developed a control-oriented model for optimization of brake-specific fuel consumption (BSFC) and NOx emissions in a direct injection (DI) diesel engine. Thanks to a statistical analysis, the model considered the most relevant parameters. They used response surface methodology (RSM) to simplify the model. They managed to predict NOx and BSFC with low errors in a couple of milliseconds [24]. Roy et al. applied a static artificial neural network to anticipate the performance and exhaust emissions of an existing one-cylinder common rail direct injection (CRDI) engine under varying exhaust gas recirculation (EGR) strategies [25]. Rezaei et al. utilized static neural networks to predict the outputs of an HCCI engine. They utilized two types of neural networks (NNs): normal feed-forward and radial basis. They remarked that feed-forward networks demanded fewer neurons (and thus was of a simpler structure) but required twice as much training time [26]. Kapusuz et al. employed neural networks to model the effect of alcohol fuel type on the performance of spark ignition engine. They used the model to present torque, brake power, and BSFC variation maps with alcohol content and also predicted the optimum methanol-to-ethanol mixture in terms of the best engine torque and power and the lowest specific fuel consumption [27]. Tosun et al. utilized linear regression and neural network for engine performance prediction. They stated that the neural network model showed better accuracy [28]. Acharya et al. made use of a neural network model to estimate the parameters of a diesel engine fueled with the mixture of diesel and mahua biodiesel in different ratios [29]. Gürgen et al. employed a neural network model to predict cyclic variations of a diesel engine with butanol diesel [30]. Yu et al. applied a NARX dynamic neural network to erect a diesel engine simulator. The simulator predicted the engine speed in a limited interval around the set-point with two inputs [31]. Di Mauro et al. used a static neural network to model and predict cycle-to-cycle variations of indicated mean effective pressure (IMEP) for an SI engine using 109 sets of data [32].

A brief outlook of some of the papers reviewed regarding the application of neural networks in engine modeling is outlined in Table 1.1.

TABLE 1.1
A Brief Outlook of Some of the Papers Reviewed Regarding Application of NNs in Engine Modeling

Year	Authors	Subject
1998	Arsie et al.	Neural network application
1998	Muller and Hemberger	Static neural network for predicting the point of 50% heat release in real time
2001	Gang et al.	One dynamic neural network for modeling a diesel engine
2002 and 2003	He and Rutland	Modeling some of the steady-state characteristics of combustion, single output networks, ensemble averaging
2003	Brahma, He and Rutland	Cylinder pressure prediction with a combination of neural network and a first-principle mathematical relation
2005	Brusca et al.	Estimation of the variable specific heat ratio for a single-zone model
2009	Samadani et al.	A static combustion model for a multi-objective optimization purpose
2011	Shivakumar and Shrinivasa Rao	Steady-state characteristics of a variable compression ratio diesel engine
2012	Uzun	Predicting steady-state characteristics of an engine
2012	Yap and Karri	Predicting steady-state emissions of an engine with three different feed-forward networks
2012	Ismail et al.	A static neural network model for a diesel engine
2012	Cay et al.	Single-output static neural networks for a methanol engine
2013	Janakiraman et al.	Some of the transient characteristics of an HCCI engine, with comparing the application of MLP & RBF, NARX, and PCA
2014	Molina et al.	A control-oriented model for optimization of BSFC and NO_x emissions in a DI diesel engine
2014	Roy et al.	Predicting steady-state characteristics with varying EGR
2015	Rezaei et al.	Predicting steady-state characteristics of an HCCI engine
2015	Kapusuz et al.	Studying the effect of alcohol fuel type on the performance of SI engine
2016	Tosun et al.	Comparing linear regression and NNs for performance prediction
2017	Acharya et al.	Modeling a biodiesel engine
2018	Gürgen et al.	Predicting cyclic variations of a diesel engine with butanol diesel
2018	Yu et al.	A two-input NARX dynamic neural network as a diesel engine simulator
2019	Di Mauro et al.	Modeling cycle-to-cycle variations of IMEP for an SI engine

As noted, the papers surveyed mostly do not model the whole engine and its transient behavior, and only dealt with some steady-state features of the combustion subsystem, such as cylinder pressure or heat release. Undoubtedly, designing a control system for an engine does invoke a complete engine model, including all engine features, from throttle body to flywheel. Only in this way, the dynamic (and transient) behavior of an engine will be captured (with factual engine inputs and outputs).

In addition, in the few cases of dynamic modeling in the papers, only some transient features were modeled via dynamic neural networks. In these papers, usually, only one output (e.g. engine speed) was predicted within a small interval.[ii]

A more advanced approach is presented in [1], where dynamic local models are utilized for global-local modeling. To do so, the engine operating space is divided into local operating points (measurement points). Local dynamic models are then constructed (using, e.g., dynamic recurrent neural networks) to represent the respective operating point. Now, a global-local dynamic model can be achieved by bilinear area interpolation of the local models (Figure 1.7). For reasons explained later, we prefer to avoid this and adopt another approach.

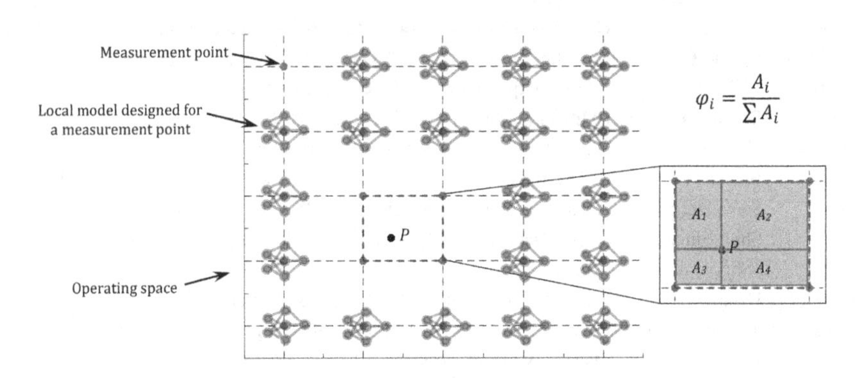

FIGURE 1.7 Dividing the engine operating space into representative measure points and assigning dynamic local models for each measure point. Note that, for the interior point P, bilinear area interpolation is utilized to achieve a global-local model (based on the four adjacent local models) [1].

There is also a facility in GT-Power software for MVM, in which the combustion chamber, in the designed one-dimensional (1-D) computational fluid dynamics (CFD) model, is replaced by a neural network. Although useful, we prefer to take a system identification approach with a more advanced neural network design, based on the data available.[iii] This will be further discussed in Chapter 2.

An old version of the modeling style to come can be found in one of the previous papers [33].[iv]

1.4 HIERARCHICAL CONTROL

As stated before, the engine control system aims to achieve the following most important objectives:

- Demanded torque with least fuel consumption
- Safe operation
- Permissible pollutant emissions

Control tasks in an SI engine could be outlined as follows [1]:

- Torque control:
  - fuel injection control
  - air/fuel ratio control
  - ignition timing control
  - combustion control
  - knock control
- Emission control:
  - exhaust recirculation control
  - catalyst control
- Idle speed control
- Coolant temperature control
- Lubrication control

In the past, engine control was basically conducted using various direct look-up tables between control inputs and outputs. However, this led to an opaque control structure with high complexity and severe interactions between variables. In 1997, Gerhardt et al. proposed torque-based hierarchical engine control [34, 35], resulting in a transparent physical-based structure. Not only did this hierarchical control structure minimize interactions, but it also separated conceptually different tasks into different layers, leading to a simpler, more effective structure.

As noted in Figure 1.4, the engine control system consists of feed-forward and feed-back control loops. As will be explained later, this combination brings both appropriate bandwidth and robustness for the system.

1.4.1 Engine Hierarchical Torque-Based Control Structure

A simple schematic of hierarchical structure is depicted in Figure 1.8. As can be seen, the structure comprises at least three different layers (inspired from [1]):

- High-level control:
  - torque demands
  - torque coordination
- Mid-level control:
  - torque conversion and feed-forward base-line signals
- Low-level control:
  - actuator feed-back control

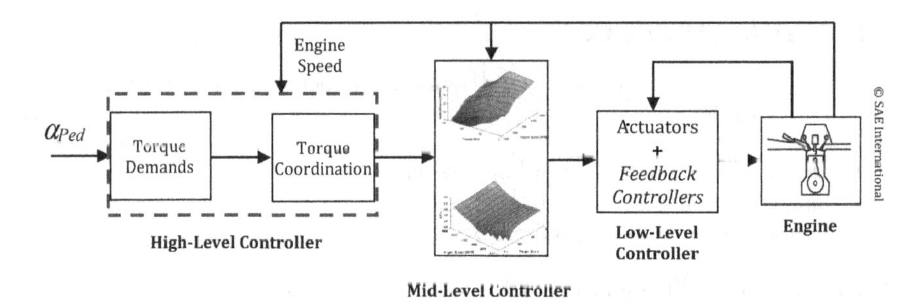

FIGURE 1.8 Engine hierarchical control structure [36].

In torque demands, all the demanded torques are computed. The main part is computed based on the gas pedal angle (driver's command) and engine speed, as the driver's demanded crude torque, for normal conditions, regardless of gear shifting, overtaking, etc. Other conditions and considerations demand their own relevant torques, such as start, idling, minimum permissible torque (to overcome losses), maximum permissible torque (to avoid component damage), transmission demands (e.g. for gear shifting), traction control demand, etc.

In torque coordination, based on the aforementioned torque demands, priories and limits, smooth operation, etc., a desired torque is determined to be realized. More information with regard to these can be found in [1].

In the mid-level control layer, based on the abovementioned desired torque, baseline signals for actuators are determined (e.g. for throttle angle, VVT, SA, etc.). In fact, this layer contains feed-forward controller maps assigning appropriate values for engine actuators in different operating conditions to fulfill the desired torque. Feed-forward controllers provide the bandwidth required for the control system. In other words, although more robust, feed-back controllers are slow and fail to provide agile real-time control, particularly in transients. Hence, feed-forward controllers are employed in this layer to first provide agile base-line actuator commands.

In the low-level control layer, the above base-line signals are modified using feed-back controllers. In fact, this layer brings robustness and more accuracy. In other words, first nearly accurate actuator commands are provided by feed-forward controllers in the mid-level layer, and subsequently, these commands are modified in the low-level layer to provide more accurate and robust control signals. Needless to say, the final control system probably shows better performance in steady states rather than transients, as feed-back controllers have better performance for low-frequency responses.

Accordingly, the engine control structure consists of several feed-forward and feed-back control loops. The principal engine control loops include the following:

- Throttle control (feed-forward)
- VVT control (feed-forward)
- Fuel injection control (feed-forward + feed-back)
- Ignition control (feed-forward + feed-back), etc.

The above loops attempt to fulfill the objectives outlined at the beginning of this section. In what follows, a short glance will be given at different control layers.

1.4.2 Mid-Level Layer

1.4.2.1 Calibration Maps

With environmental standards becoming more and more stringent, reducing fuel consumption while meeting pollutant emission limits is an increasing challenge in the automobile industry. Even for hybrid electric vehicles, a significant portion of torque generation is accomplished by the combustion engine, mostly outside of sweet spot, and in ordinary part-load points.[v] These all demand proper engine control design.

In the light of the previously mentioned arguments, a key section of an engine hierarchical control structure is the mid-level layer, which comprises feed-forward controller maps assigning appropriate values for engine actuators in different operating conditions, based on the demanded torque from high-level controller. These feed-forward maps (aka, calibration maps) are derived based on the two aforesaid objectives, namely, minimizing fuel consumption while meeting emission standard limits, along with maintaining the engine work within safe and preferable margins. This is basically a constraint optimization problem. The calibration maps provide base-line signals for actuators, which could be modified by feed-back controllers in the low-level control (Figure 1.8).

Owing to the problem complexity, together with the conflicting effects of different control inputs on the above objectives (and of course, the conflicting behavior of the objectives, themselves), and also the abundance of the engine operating points, manual optimization at a test bench would be highly time-consuming, costly, and complicated. Consequently, a computer *model-based calibration* is required [1, 2], in order to first find the optimal operating points (or their approximate values) by computer, and consequently, save in time and expenses demanded for final calibration at the test bench.

Model-based calibration procedure, as the name suggests, involves two steps: modeling and optimization. The model is supposed to accurately predict the impacts of control inputs on system outputs, preferably in real time. Afterward, the optimization strategy comes into play to find the appropriate engine inputs for optimal operating points and to yield the calibration maps. A simple schematic of the inputs and outputs of an SI engine is portrayed in Figure 1.5. As will be thoroughly explained later, according to emission standards (such as New European Driving Cycle [NEDC] or World Harmonized Light-duty Vehicles Test Procedure [WLTP], etc.), the amounts of NOx, CO, UHC (unburnt hydrocarbon), and CO_2 emissions in a special driving cycle shall not exceed specified values.

With regard to modeling, various approaches have been adopted in the literature. Apart from applications of neural networks explored in the previous section, some like Rask and Sellnau, Zhao and Xu, and De Bellis have developed GT-Power software models as data-set creators for calibration [37–39]. In spite of higher accuracy (due to 1-D CFD modeling), simulations become overwhelmingly time-consuming,[vi] and not suitable for an optimization task with hundreds of points, to say the least. Accordingly, control-oriented models with fast response are required. Another example of using special software can be found in [40]. Grasreiner et al. suggested a phenomenological combustion model considering cylinder thermodynamics, combustion-relevant turbulence, and ignition delay to predict some in-cylinder characteristics of an SI engine. They employed GT-Suite software for all the simulations [41].

Langouët et al. utilized local polynomial models [42, 43]. Isermann and Sequenz employed global-local polynomial models (as a combination of local polynomial models) [44]. In these cases, D-optimal approach was used for the design of experiments. Park et al. exploited a regression model based on the radial basis function [45]. Millo et al. utilized Gaussian process models [46]. Some have made use of other (fully black-box) neural networks for engine modeling. Hafner and Isermann, in a notable work, applied LOLIMOT algorithm for modeling [47]. Atkinson and Mott exploited dynamic neural networks for engine modeling [48]. Samadani et al. applied static neural networks for the steady-state prediction of a diesel engine [17]. Schaberg and Atkinson, following the Atkinson's previous work, [49], used neural networks for engine modeling [50]. As a matter of fact, most of these models are local models, based on which, as explained in [51], the resulting optimization maps might need harsh smoothening (due to switching from one local model to another), degrading the accuracy of the calibration. The reason why local modeling is mostly employed in the literature and industrial applications is the fact that prevalent global models lack accuracy. As will be discussed in this study, an engine *global* model with high accuracy will be presented.

As regards optimization, different techniques and methodologies have been utilized in the literature. In terms of optimization techniques, gradient-based methods are rarely applied (like in [50], as they did not consider the catalytic converter). It will be explained later that the optimization problem is prone to be non-convex, and as a result, global meta-heuristic approaches need to be employed. Genetic algorithm (GA) and particle swarm optimization (PSO) are the most commonly used optimization techniques. Tayarani et al. comprehensively reviewed the applications of various meta-heuristic algorithms in engine optimization, from GA and differential evolution (DE) to particle swarm and ant colony (AC), and hybrid methods [52]. They maintained that more than the optimization technique, it is in fact the modeling quality that matters, in engine calibration.

In terms of the optimization methodology, some have utilized online optimization (and or modeling). Malikopoulos et al. employed the Markov Decision

Process and online learning of engine dynamics. They deemed the system behavior as a random process and exploited reinforcement learning to solve the optimization (sequential decision-making) problem [53]. They only optimized SA to maximize the produced torque, yet, it might actually contradict emission requirements. In another study, they made use of online decentralized learning control scheme to optimize the injection timing and variable-geometry turbocharger (VGT) vane position for a diesel engine [54]. Boes et al. developed a self-adaptive multi-agent system, named ESCHER, for model-free optimization of an internal combustion engine. This method was based on the cooperative self-organization of agents and comprised observing the process, representing criteria, analyzing the state of the environment, and selecting the adequate action. The agent capable of meeting the relevant criterion with a lower value would take action; but, if no adequate action was available, it would self-organize itself [55]. Nevertheless, this method demands plenty of time for learning each point. Wong et al. suggested an online point-by-point approach for the calibration of an SI engine. An extreme learning machine (ELM) was applied for modeling. Starting from one point and moving to neighboring points, the system was gradually optimized using the particle swarm method. In each step, the model modified in the last step was employed as the initial basis for the new optimization. By means of a proposed method for the design of experiments, the model was updated and a new optimization was carried out [56]. Although interesting and innovative, these approaches might not be real time when implemented in an online real-engine controller. Furthermore, supposing pre-optimization before real application, a huge number of tests and/or simulations might be demanded.

Multi-objective optimization is employed as for diesel engines, or when a catalytic converter is not modeled. Langouët et al. adopted multi-objective covariance matrix adaptation evolution strategy upon NEDC for the calibration of a diesel engine. They asserted this algorithm to be of a faster response. They regarded minimizing NOx emission as the primary objective [42, 43]. Samadani et al. exploited NSGA-II to derive Pareto front for NOx, Soot, and IMEP. In the end, they deemed NOx minimization as the main objective [17].

As mentioned before, as for light-duty vehicles, standard driving cycles' emission criteria (for each type of pollutant emission) are *not* stated for each specific operating point, but rather, with respect to gram per kilometer, for the *complete* driving cycle. This necessitates the optimization procedure to be based on the driving cycle. Notable examples of NEDC-based calibration by Hafner and Isermann, and Iserman and Sequenz can be found in [44, 47]. It also demands modeling catalytic converter, which is hardly considered in the literature.

The resulting optimization maps are normally uneven and bumpy, and demand smoothening to be practical for actuator control. This roughness is partly due to local modeling (which as previously mentioned; it would be avoided in this study). Moreover, meta-heuristic optimization, although global, is sub-optimal and approximate rather than accurate and causes fluctuations in the maps. Langouët et al. [42, 43] and Iserman and Sequenz [44] made use of the LOLIMOT algorithm for map smoothening. Millo et al. applied a two-step optimization procedure, including random calibration, and then, GA. They maintained that this approach is able to make the calibration maps smoother. Arya et al. adopted the Gaussian curvature method and selected the smoothest maps amongst half a million different combinations of local-model-based maps [51]. Nikzadfar and Shamekhi utilized the previously optimized points in the vicinity as constraints to obtain smoother maps [57]. As will be seen, a simple but effective method will be employed in this study.

In terms of optimization target, different strategies have been adopted in the literature. Above all, the constraint optimization target could be either maximizing the produced torque or minimizing fuel consumption. Some have attempted to maximize the produced torque [53, 58] without considering the emission constraints. On the other hand, some like Nikzadfar and Shamekhi separated part-load and full-load modes and assigned a different target for each. They assigned minimizing fuel consumption for part-load mode, and torque maximization for full-load mode, for a diesel engine [57].

1.4.2.2 Air Mass Flow Observer

As mentioned in the beginning of this section, fuel injection subsystem is a combination of feed-forward and feed-back controllers. The feed-forward controller is in fact an in-cylinder air mass flow observer. In other words, a dynamic model of the intake manifold is usually utilized to estimate the air mass flow into the cylinder, and the amount is then multiplied by the stoichiometric fuel-air ratio to yield the demanded fuel flow rate. This feed-forward controller signal is then modified by the feed-back controller. For the sake of better understanding and coherence, this part is mainly discussed in Chapter 4.

1.4.3 Low-Level Layer

1.4.3.1 Fuel Injection Feed-Back Control

Fuel injection control is one of the most crucial control loops in SI engines. Particularly for port fuel injection SI engines (Figure 1.9), which still constitute a substantial portion of vehicle engines in the world, proper fuel injection control has always been a matter of great challenge.

Basically, for a three-way catalytic converter to perform appropriately and meet the emission standards, the in-cylinder air-fuel ratio (AFR) needs to be kept within a narrowband around stoichiometry,[vii] called the lambda window. Accordingly, any flaw in this control system may cause pollutant emissions to exceed standard limits.

To enhance the fuel injection control and bring robustness, an AFR sensor (called λ-sensor) is installed downstream of the combustion chamber, to form a feed-back control loop (Figure 1.9). Nonetheless, the current AFR value of the flow into cylinder is relayed back with a delay, after its combustion. In the automotive industry, since the λ-sensor feedback is delayed (and renders the control

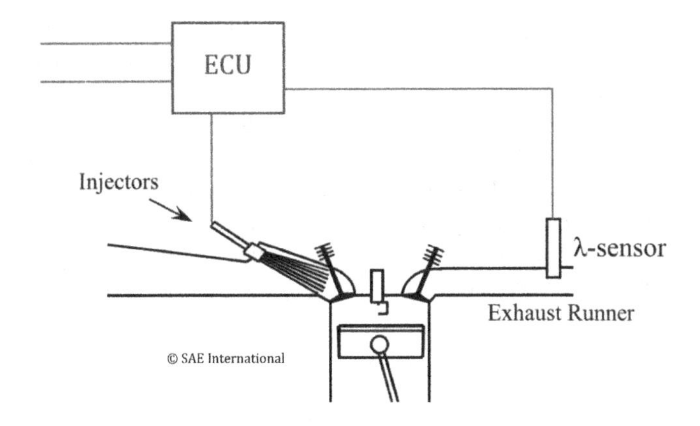

FIGURE 1.9 Fuel injection feed-back control subsystem [59].

system not agile), a pure feed-back control system is avoided, and a combination of feed-forward and feed-back controllers is applied, in order to achieve better accuracy, bandwidth, and robustness (Figure 1.10). The duty of the feed-forward controller is to instantly provide an estimation of the amount of fuel required for injection into intake ports (in port fuel injection engines). In other words, this feed-forward controller provides the bandwidth and the agility required. This is realized using an in-cylinder air mass flow rate observer.[viii] That is, a dynamic model of intake manifold is usually employed to estimate the air mass flow rate into cylinders, and the amount is then multiplied by the stoichiometric fuel-air ratio to yield the demanded fuel flow rate:

$$\dot{m}_{\text{fuel}} = \dot{m}_{\text{air}} \cdot \sigma, \tag{1.1}$$

where σ is the stoichiometric fuel-air ratio. Although not completely robust (and thus, accurate), this controller provides the bandwidth required, particularly in transient conditions.

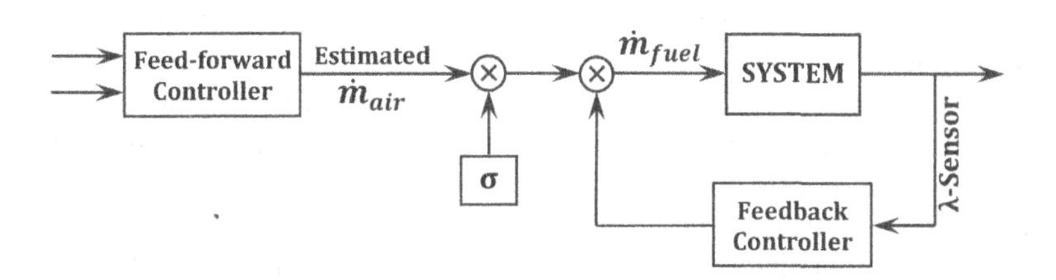

FIGURE 1.10 The combination of feed-forward and feed-back controllers.

The feed-back controller, on the other hand, is applied to modify the set point estimated by the feed-forward controller, based on the λ-sensor signal fed back from downstream of the combustion chamber. Normally, a proportional–integral

(PI) controller is employed in the feed-back loop. Although delayed and of lower bandwidth, this controller provides robustness and more accuracy, particularly, for steady-state conditions.

Nonetheless, the aforementioned structure fails to adequately address transient conditions (Figure 1.11). It shall be noted that 40–50% of the NEDC occurs in transients [1]. This escalates when it comes to newer driving cycles, such as FTP-75 (2008) and particularly WLTP (2015). Furthermore, the system parameters vary in different operating conditions, and fixed gain controllers could not yield satisfactory performance.

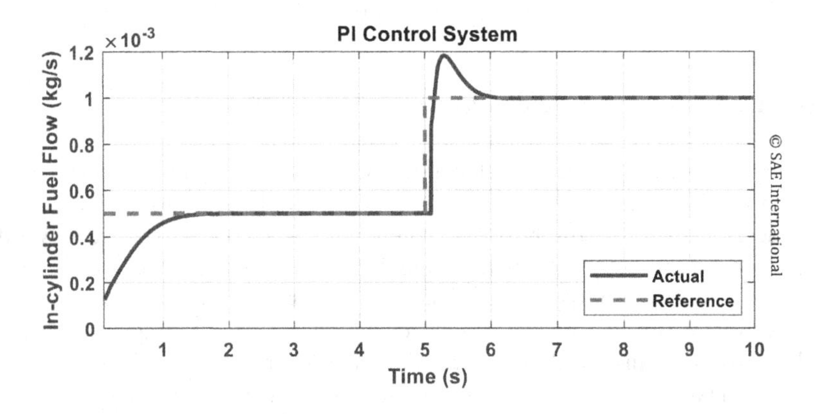

FIGURE 1.11 The transient behavior of an ordinary PI fuel control system [59].

Some papers have attempted to enhance the feed-forward observer. Stotsky and Kolmanovsky suggested an in-cylinder flow rate observer consisting of three interconnected observers for better flow estimation in steady and transient states. It comprised a throttle flow estimator, an intake pressure estimator, and a third observer for online correction [60]. The observer still had small errors for transients. Arsie et al. proposed an adaptive estimator based on the extended Kalman filter for fuel film dynamic in intake manifold. The estimator was a combination of a mean-value feed-forward observer and an extended Kalman filter using the λ-sensor signal [61]. It had a decent performance, with fluctuations in the beginning seconds of operation. Albrecht et al. designed a 1-D model of a turbo-charged SI engine in AMESim software, from which data was acquired and stored in two neural network models as estimators of air mass flow rate into cylinders [62]. Nonetheless, these types of observers fail to be robust for aging. Le Solliec et al. made an effort to resolve this problem by adding a Kalman filter in a closed-loop form to the neural-based observers [63]. Meng et al. employed the extended Kalman filter to enhance the transient estimation of the intake air mass flow rate and instantaneous cylinder charge [64]. Wu and Tafreshi suggested an observer-based internal model controller to cope with system parameter uncertainty and delay. They applied Padé first-order approach, with a state observer and an integrator [65].

Some works have focused on feed-back control, yet ignored feed-forward observation. Tang et al. designed two nonlinear centralized controllers for SI engines. The first was an adaptive controller, and the second was based on learning control. The control outputs comprised fuel injection, throttle angle, and ignition timing [66]. Nevertheless, this centralized control demands set-points for manifold pressure and engine speed. Tang et al., in another study, compared sliding mode with globally linearizing control (GLC) for AFR and EGR control [67].

Some others have applied online or offline adaptive algorithms to address parameter variations and to improve transient behavior. Liu et al. used adaptive critic learning for calibration and control of torque and AFR of an automobile engine. They constructed neural network models and controllers using approximate dynamic programming for optimal performance. Adaptive critic learning consists of an action network with an adaptive critic for correction [68]. Muske et al. designed a predictive controller based on adaptive linear state space for air-fuel control. Model parameters with variable delay (for predictive control) were adapted based on instantaneous engine speed and load by means of offline look-up tables. The feed-back control modified the conventional feed-forward controller [69]. Ebrahimi et al. proposed a parameter-varying proportional–integral–derivative (PID) with a compensator for AFR control. They employed Padé approximation for the system delay and attempted to resolve the internal instability of the resulting non-minimum phase dynamic system with a compensator [70]. Efimov et al. designed two controllers switching to each other, with a supervisor. The first was a robust controller providing global stability but with mediocre performance. The second was an adaptive controller aiming to enhance the transient response and was activated by supervising the performance of the first controller [71]. Zhou et al. suggested adaptive disturbance rejection control based on adaptive extended state observer and utilized it for AFR control [72]. Kumar and Shen employed a predictive controller for AFR control of a turbo-charged gasoline engine. They made use of a Kalman filter to estimate the in-cylinder flow and residual gases. The model parameters had been stored by offline experiments [73]. Song et al. designed a disturbance rejection AFR controller considering transport delay. An extended state predictor observer and a recursive least square (RLS) observer were utilized to correct a physics-based AFR model with exhaust gas transport delay, and the system was then controlled as a first-order linear system [74]. Na et al. addressed AFR control as a tracking problem and suggested the application of existing offline gain-scheduling PIDs. Using Lyapunov's approach, they estimated the lumped engine's unknown dynamic [75]. However, their approach was based on the assumption of accurately measuring in-cylinder air-mass flow rate, which is optimistic.

There are also some applications of sliding mode for AFR control. Wagner et al. applied and compared two nonlinear control structures of the sliding mode and back-stepping instead of the conventional feed-forward and feed-back combination [76]. Nonetheless, they assumed the subsystem without delay and with constant parameters. Piltan et al. utilized sliding-mode control for AFR control [77]. Wu and Tafreshi made use of fuzzy sliding mode for AFR control [78].

As mentioned earlier, unlike the practice adopted in some of the works above, the combination of feed-forward and feed-back control is of great significance in achieving both the required bandwidth and robustness for the system. Moreover, fixed-gain controllers fail to adequately deal with parameter variations and transient phases, leading to the choice of *adaptive* controllers.

Adaptive controllers applied in the literature, as reviewed, are either online or offline. The issue with online adaptive controllers is their high computational burden on the ECU, especially at high engine speeds. That is, as the engine speed rises, the computation time available for ECU drops, and online adaptive control may not be feasible at high speeds. Offline adaptive controllers (whose varying parameters are tuned and stored offline), in spite of faster response, are not robust, particularly as for system aging and uncertainties of the feed-forward controller.

1.4.3.2 Spark Advance Feed-Back Control

As previously mentioned, SA control is also a combination of feed-forward and feed-back controllers. The feed-forward controller is located in the mid-level layer, yet the feed-back controller lies in the low-level layer. Ideally, for optimal SA feed-back control, the cylinder ought to be equipped with an in-cylinder pressure sensor. However, this entails considerably more expenses and is rarely utilized.

Knock avoidance is one of the major requirements of SA control. In fact, the knock sensor signal is fed back to the ECU for safe ignition to be guaranteed. However, safer ignition means retarded spark timings, which per se, contradicts lower fuel consumption. As a result, engine operation is usually maintained close to knock-prone crank angles.

Knock is detected using a bandpass filter separating knock pressure waves from cylinder pressure. Knock could be resolved by delaying ignition timing or reducing throttle angle. As the response time of the two approaches differ, a combination of them could be utilized. Figure 1.12 portrays the respective block diagram [2].

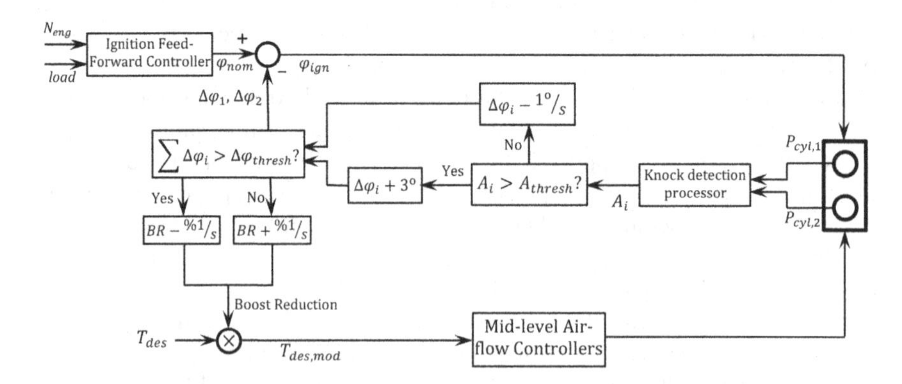

FIGURE 1.12 SA control block diagram [2].

Of all engine control inputs, SA has the fastest impact on engine performance as it affects the very upcoming engine cycle. Thus, delaying the ignition timing could readily eliminate knock occurrence. However, the downsides are torque reduction and exhaust gas temperature increase, both resulting from the delay in ignition timing. The modified spark angle is defined as

$$\varphi_{ign} = \varphi_{nom} - \Delta\varphi, \tag{1.2}$$

where φ_{nom} is the nominal ignition timing determined by the feed-forward controller in the mid-level layer, and $\Delta\varphi$ is the spark angle correction. Note that φ is positive, and $\Delta\varphi$ is restricted to positive values. At any engine cycle for the cylinder i, the maximum (bandpass-filtered) pressure signal amplitude A_i is calculated and compared with the threshold value A_{Thresh}. If the former exceeds, the ignition timing is retarded by 3° CA. Afterward, as long as $A_i < A_{Thresh}$ holds, ignition angle is advanced at the rate of 1° CA/s. Should the ignition angle command from the feed-forward controller cause knock, a "limit cycle" would occur [2].

Load reduction,[ix] on the other hand, is of slower response and is adopted once delaying the ignition timing cannot resolve knocking alone; for instance, when further retarding the ignition angle leads to overly high exhaust gas temperature, and thus, component damage. Accordingly, once sum of the ignition delays in all cylinders exceeds, boost reduction is undertaken at the rate of 1% per second. On the other hand, load is boosted at the same rate providing that the sum is lower than the threshold.

There is a similar additional facility in some ECUs that utilizes the faster impact of SA on the produced torque (compared to throttle angle variation), in a separate command path. In this practice, when a fast torque change is demanded (e.g. when automatic gear shifting or demands from traction control system) this path is triggered to swiftly change torque to the desired value through SA adjustment.

SA control has been long discussed in the literature. Powell explored the usage of in-cylinder pressure sensor for SA control, AFR, and misfire detection. He claimed that the maximum pressure for optimal SA is constant for different engine operating conditions, and could be employed for SA feed-back control. Additionally, unlike the lambda sensor, AFR estimation using the pressure sensor does not require a fully warmed-up engine [79]. Auzins et al. proposed the usage of ion current in the combustion chamber for misfire detection, knock, and engine control. They indicated that misfire and knock detection via ion current is simpler compared to ordinary approaches [80]. Sellnau et al. examined the application of the pressure sensor for closed-loop control of SA and EGR. They maintained that setting PRM10 parameter[x] to 0.55 with SA regulation, in addition to yielding optimal efficiency, allows NOx and UHC emissions to be minimized with EGR control [81]. Haskara et al. presented a retard limit for ignition timing control using in-cylinder ion-current signals. Ion current was employed to yield a criterion for combustion quality, and a retard limit was presented [82]. Huang and Chen employed offline surf-fitting to achieve ignition timing map for an SI engine. They made use of 256 operating points derived from the engine test to obtain the map [83].

Shamekhi and Ghaffari employed in-cylinder ion current for SA feed-back control. The resulting fuzzy controller was capable of locating the maximum pressure angle in the desired position [84]. Lezius et al. exploited the in-cylinder pressure signal for SA control. They used a high-pass filter for knock detection and a low-pass filter for the distance-to-knock estimation. They attempted to set the ignition timing as close as possible to the knock limit [85]. Tang et al. proposed two centralized multi-input multi-output (MIMO) control structures so as to regulate fuel injection, throttle angle, and SA. The first was an adaptive controller, and the second was based on learning control. The results were compared with GLC [66]. It shall be noted that the application of centralized control demands set-points for manifold pressure and engine speed, which is not so logical.

Using model predictive control (MPC), Behrendt et al. attempted to regulate the output torque and speed of an SI engine. They stated that the downside of an explicit approach for multi-input systems is the huge memory required. Instead of direct throttle angle and SA control, their proposed MPC would determine the set points for the air path flow into the cylinder and the torque changes caused by ignition timing [86].

In another study regarding the application of in-cylinder ion current, Malaczynski et al. examined knock detection and flame stability [87].

Tamaki et al. suggested an online approach to obtain the ignition timing map of an SI engine. The method was based on online learning of the ignition timing of maximum power using a pressure sensor [88]. Nonetheless, apart from the overwhelming amount of time required for learning, it shall be noted that the objective of ignition timing control is not necessarily maximum power since it may contradict emission limits. Jamali et al. applied proportional-integral-plus (PIP) control to regulate the produced torque, AFR, and knocking of an SI engine. The inputs included throttle angle, SA, and EGR valve position. The linear model for controller design was derived via a refined instrumental variable. The control gain matrix for the PIP control design was obtained using linear quadratic (LQ) optimal design [89]. Nevertheless, the control design is not sophisticated enough for appropriate engine control. Zhang et al. employed online self-optimizing control of SA to maximize the indicated fuel conversion efficiency. To address the stochastic optimization problem (with dithers), they utilized a gradient descent-based dither-free optimization algorithm [90]. Yet, again, it shall be noted that maximum efficiency control does not necessarily meet emission regulations. Maldonado and Stefanopoulou proposed a stochastic likelihood-based misfire controller for the engine to operate close to the misfire limit, and to cope with combustion variability in high EGR operation. To generate a continuous model for simulation, they applied third-order polynomial fitting [91]. Lee et al. examined the effects of varying ignition timing and excess air ratio on the combustion and NOx emissions of a hydrogen-fueled spark ignition engine under part-load conditions [92].

Table 1.2 summarizes the papers reviewed regarding SA feed-back control.

TABLE 1.2
A Summary of the Papers Reviewed Regarding SA Feed-Back Control

Year	Authors	Subject
1993	Powell	Usage of in-cylinder pressure sensor for SA control, AFR, and misfire detection
1995	Auzins et al.	Application of ion current in the combustion chamber for misfire detection, knock and engine control
2000	Sellnau et al.	Application of pressure sensor for closed-loop control of SA and EGR.
2004	Haskara et al.	Presented a ritard limit for ignition timing control using in-cylinder ion-current signals.
2006	Huang and Chen	Offline surf-fitting to achieve ignition timing map for an SI engine.
2007	Shamekhi and Ghaffari	In-cylinder ion current for SA to design a fuzzy feed-back controller
2007	Lezius et al.	In-cylinder pressure signal for SA control, with a high-pass filter for knock detection, and a low-pass filter for the distance-to-knock estimation.
2009	Tang et al.	Two centralized multi-input MIMO control structures so as to regulate fuel injection, throttle angle, and SA (including an adaptive controller and a learning controller)
2011	Behrendt et al.	Application of MPC to regulate output torque and speed of an SI engine, with set points for the air path flow into the cylinder and the torque changes caused by ignition timing
2013	Malaczynski et al.	Knock detection and flame stability using in-cylinder ion current
2014	Tamaki et al.	Ignition timing map using online learning of the ignition timing of maximum power using a pressure sensor
2015	Jamali et al.	Application of PIP control to regulate the produced torque, AFR, and knocking
2017	Zhang et al.	Online self-optimizing control of SA to maximize the indicated fuel conversion efficiency with gradient descent-based dither-free optimization algorithm
2018	Maldonado and Stefanopoulou	Stochastic likelihood-based misfire controller for the engine to operate close to misfire limit, and to cope with combustion variability in high EGR operation
2019	Lee et al.	Effects of varying ignition timing and excess air ratio on the combustion and NO_x emissions of a hydrogen-fueled spark ignition engine under part-load conditions

1.4.4 High-Level Layer: Torque Control

Providing the demanded torque is probably the primary duty of an ECU. The demanded torque is interpreted based on the driver's command, i.e. gas pedal angle, at the respective engine speed and operating conditions, and some other design considerations. As for hybrid electric vehicles, this torque is demanded (from engine) by EMS, based on some extra factors, such as the state of charge, operating mode, etc. The ECU shall adjust the engine control inputs (e.g. SA, throttle angle, VVT, fuel injection, etc.) such that the demanded torque is realized within a sensible duration of time.

As mentioned before, (what we called) high-level layer, in fact, consists of (at least) two sections, namely torque demands and torque coordination (Figure 1.8). The output of this layer is a desired torque (or equivalently, a desired in-cylinder air mass flow rate) demand. This demand is converted to actuator base-line set points in the mid-level layer. On the whole, this combination of feed-forward controllers in high-level and mid-level layers yields responses of decent bandwidth.

In conventional ECUs, the final demanded torque is put through (for instance) a first-order filter (before submitting to the mid-level layer) to avoid a sudden rise or fall and to yield a gentle variation. Although of proper bandwidth and fast response, this feed-forward torque control approach is not robust (and probably, not accurate). It is also not optimal during transients.

Different approaches have been suggested for torque control in the literature. Owing to high cost, vehicles are not equipped with torque sensor, and hence, some articles have focused on designing torque observers. Some examples could be found in [93–102].

Articles regarding torque control mostly focus on throttle angle manipulation. Chamaillard et al. suggested a robust controller for engine torque based on throttle angle variation [103]. It was assumed that SA would be later computed based on throttle angle. Liu et al. utilized adaptive critic learning for calibration and control of torque, and AFR of an automotive engine. They constructed the neural network models and controllers using approximate dynamic programming for optimal performance. Adaptive critic learning comprises an action network with an adaptive critic to correct it [68]. Huang et al. made use of neural sliding-mode control for engine torque via manipulating throttle angle. They incorporated neural networks into a sliding-mode controller to overcome the chattering thereof. They employed two parallel networks one as equivalent control and the other as corrective control [104]. Behrendt et al. exploited MPC for torque and speed regulation of an SI engine. The controller would determine input references for throttle angle and SA [86]. Jamali et al. employed PIP control for torque, AFR, and knock in an SI engine. Manipulated inputs consisted of throttle angle, SA, and EGR valve. The linear model for controller design was identified using a refined instrumental variable, and control matrices were derived via the LQ optimal method [89]. Zhou et al. proposed a torque controller for a turbo-charged SI engine with air-path regulation. Air-path control comprised manipulations of wastegate and throttle angle. The controllers contained predefined maps and PID controllers. Nonetheless, their work had quite a few simplifications [72]. Bao and Kong utilized generalized predictive control for the transmission control unit. They intended to address the problems concerning time delay and parameter perturbation of engine torque control in gear shifting of dual-clutch transmissions [105]. Anjum et al. suggested a dual-loop control strategy to track the desired speed profile for engine torque management, with one-loop manipulating throttle angle and the other regulating fuel injection [106].

In one of our papers in 2020, nonlinear multi-parametric MPC was employed for engine idle speed control. Online nonlinear model predictive control (NMPC) is too time consuming, and thus, not applicable for a real engine control system. Therefore, a global map was derived from *offline* solution of NMPC for a wide range of operating points with the receding horizon. The control system, in the end, was coupled with a low-gain PI controller to be able to cope with disturbances [107].

The articles reviewed, typically ignore the hierarchical control structure, and merely focus on feed-back controllers. It shall be noted that the engine is a complicated MIMO system and ignoring the hierarchical structure results in high interactions. Apart from this, totally ignoring feed-forward controllers and merely relying on feed-back control is highly unlikely to provide real-time responses and the bandwidth required. It needs to be noted that the system is delayed and also the computation time available for ECU in each engine cycle is extremely short.[xi] Additionally, feed-back control is of lower bandwidth compared to feed-forward control. This problem escalates as the feed-back controller becomes more complicated, further degrading the agility of the control system. Consequently, mere feed-back controllers (especially complicated ones) are not practical, and the hierarchical structure, with the combination of feed-forward and feed-back controllers needs to be applied.

On top of these, feed-back controllers (which are of lower bandwidth) basically concern steady-state responses (i.e. low-frequency response) and may fail to account for transients (which are high-frequency responses). As mentioned, a large proportion of standard driving cycles (especially the newer ones) occur in transient states. This escalates when it comes to hybrid electric vehicles. As a matter of fact, in hybrid electric vehicles, the engine has to be frequently started off, and loaded up to desired torques, raising the importance of transient operation control. In other words, the transition path to the desired torque significantly affects the overall amount of fuel consumed[xii] (and even the overall pollutant emission produced).

In this book, an alternative approach for torque control will be presented, that is capable of transient optimal control.

1.5 ENGINE PROPERTIES FOR CASE STUDY AND DATA ACQUISITION

Control-oriented modeling is based on experimental data acquired from a real engine test. For the following reasons, it *might* be necessary to combine these data with data acquired from a CFD model accurately designed, calibrated, and validated based on the experimental data:

- The number of experimental data sets available may not be enough for identifying the (complicated) system, or they might not cover the whole engine operating regions.

- The experimental data provided by an automotive industrial unit are often acquired from an engine connected to the ECU, and in this way, the system outputs are only obtained from favorable controlled inputs. It shall be noted that the final control-oriented model (which will be designed based on the acquired data) must be able to accurately predict the system outputs, for both favorable and unfavorable inputs.
- As for each operating point, some intermediate parameters, necessary for designing the model, may not be found in the experimental test tables (e.g. mass flow rate, when only a manifold pressure sensor is available).

Hence, the experimental data table (acquired from a dynamometer, Figure 1.13) could be augmented with data from an accurately validated CFD model (e.g. GT-Power software[xiii] model), to attain a comprehensive pattern table, covering all the engine favorable and unfavorable operating points, to train neural networks. More information in this regard can be found in [108]. We call this combined data, "Software-Experimental data." While demanding and time-consuming, this step has a significant impact on system modeling. Needless to say, should all the necessary experimental data be available, no CFD-model data acquisition is demanded.

FIGURE 1.13 Engine dynamometer.

The engine available for the case study is naturally aspirated. Some of the specifications of the in-line four-cylinder SI engine with VVT employed in this study are outlined in Table 1.3. Table 1.4 presents the fuel properties. Specifications of the three-way catalytic converter are outlined in Table 1.5. Figure 1.14 depicts the GT-Power model designed for supplementary data acquisition.

TABLE 1.3
Engine Specifications

Specification	Value
No. of cylinders	4
No. of valves	16
Maximum power	113 hp @ 6000 rpm
Maximum torque	155 N·m @ 3500–4500 rpm
Displacement volume	1645 cc
Compression ratio	11.05
Bore	78.6 mm
Stroke	85 mm
Connecting rod length	134.5 mm
Throttle diameter	51.8 mm
Maximum valve lift	8.3 mm
Valve lash	0.1 mm

TABLE 1.4
Fuel Properties

Property	Value
Molecular weight	115 mole/g
Carbon (% mass)	87.7
Hydrogen (% mass)	12.2
Stoichiometric ratio	14.21
Lower heating value (LHV)	43.84

TABLE 1.5
TW Catalytic Converter Properties

Property	Value
Frontal area	12,449 mm^2
Chamber length	136 mm
% Area open to flow	71
Cell density	600 in^{-2}

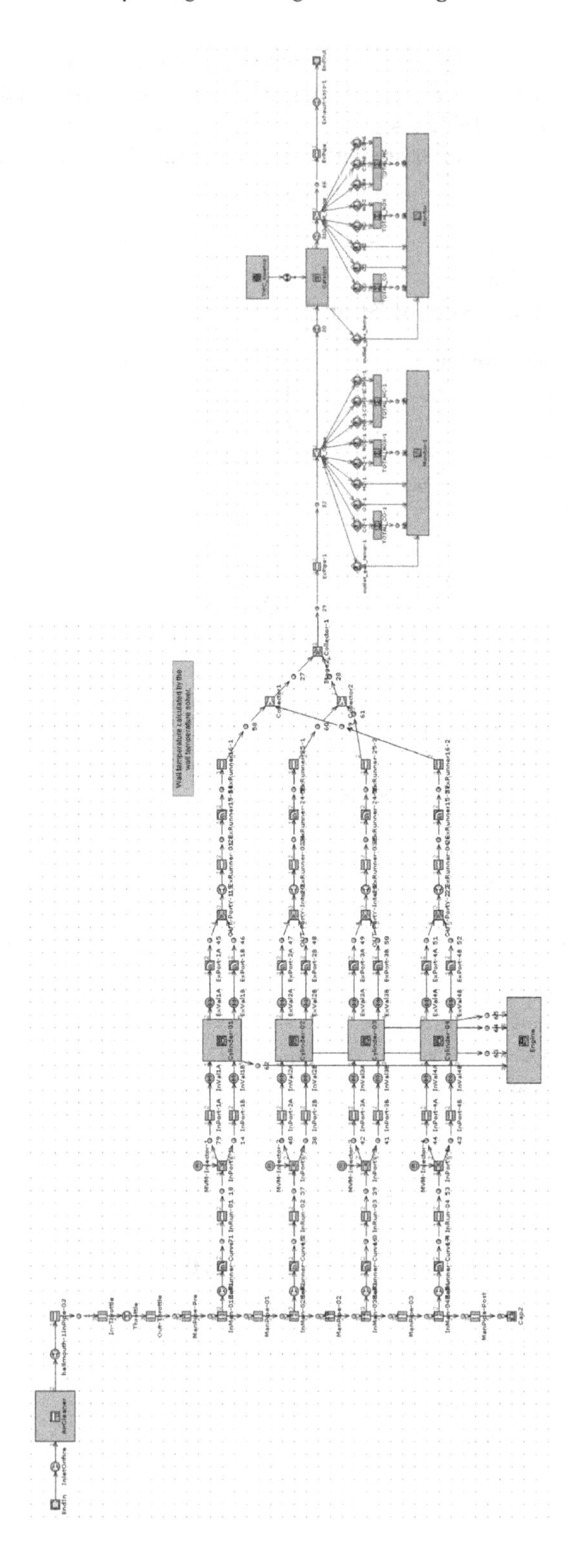

FIGURE 1.14 The final GT-Power model designed for supplementary data acquisition. The model must be carefully calibrated and validated based on experimental data.

1.6 OVERVIEW, ORGANIZATION, AND STRUCTURE GUIDELINE

As noted, with the increasing tendency toward lower emissions and also hybrid vehicles, besides steady-state performance, engine transients are becoming more and more significant.

More importantly, control engineering, particularly during the past decade, has been seeking optimality for nonlinear systems. This will be heeded in this study.

The aim of this book is to present step-by-step MiL control system design, through a case study for a port fuel injection SI engine, with focus on optimal control for both transient and steady-state behaviors. The methodology presented could be extended and employed for other combustion engines and standards.

In the chapters to come, each of the topics proposed previously are to be described in detail.

In Chapter 2, control-oriented modeling is presented. The model in this study is a global gray-box model capable of accurately simulating engine transient and steady-state behavior in real time.

Chapter 3 is dedicated to mid-level control design and engine calibration. Based on the global model, a cycle-based calibration is conducted.

Fuel injection feed-back control is examined in Chapter 4, as low-level control. To properly address transients, adaptive MPC is utilized. With regard to SA feed-back control, since the engine in hand does not contain in-cylinder pressure sensor, only the conventional knock avoidance feed-back controller (described in Section 1.4.3.2) is employed.

Chapter 5 explores optimal nonlinear torque control in transients. The methodology presented could be of paramount significance for transients and hybrid vehicles.

NOTES

i. Black-box models, unlike white-box models, do not get involved in solving the real equations governing a system, but rather try to estimate the outputs of a system for the given inputs, based on prior knowledge given to them, without any physical insight. They include a wide variety of methods, from classical simple look-up tables to sophisticated machine learning techniques.

ii. It will be explained that this can be due to burdening this extremely complex task on one dynamic neural network.

iii. Regardless of whether the data is acquired from experiments or both experiments and software.

iv. In this paper, we achieved a dynamic, real-time, and accurate model. The model, however, lacks some indispensable subsystems and some inputs and outputs necessary for a practical control-oriented model or simulator and will be completed here.

v. Particularly as for parallel hybrid electric vehicles and to some extent for series–parallel hybrid electric vehicles.

vi. Even though Zhao and Xu utilized four workstations in cluster and enhanced the process by different approaches, optimization of each point took about 5 hours!

vii. In rich air-fuel mixtures, the catalytic convertor fails to eliminate unburned hydrocarbons, and in lean mixtures, it fails to abate nitric oxides [2]. The air–fuel ratio must stay within $\lambda = 1 \pm 0.005$ [1].

viii. Sometimes, a simple compensator is also added to account for wall-wetting effects.
ix. Load means the amount of torque desired to be produced by the engine. It is directly related to (desired) in-cylinder mass flow rate and is foremost controlled by throttle angle.
x. PRM10 is defined as PRM10 = [(Pressure Ratio at 10° ATDC) – 1]/[(Final Pressure Ratio) – 1]; in which, ATDC stands for after top dead center, and Pressure Ratio is defined as the fired pressure divided by the motored pressure at ϑ° ATDC.
xi. At most, between 0.2–0.02 s, depending on the engine speed.
xii. As a matter of fact, there are studies in the literature aiming to improve transient behavior of the control system, by considering the low-level or mid-level controllers. However, as noted, the influence of the high-level controller is more substantial in transients, and this is where we will construct our controller.
xiii. GT-Power is a 1-D CFD commercial software, broadly used in engine design and simulation.

REFERENCES

1. Isermann, R., *Engine Modeling and Control: Modeling and Electronic Management of Internal Combustion Engines.* 2014, Berlin, Germany: Springer.
2. Guzzella, L. and C. Onder, *Introduction to Modeling and Control of Internal Combustion Engine Systems.* 2nd ed. 2010, Berlin, Germany: Springer.
3. Isermann, R., J. Schaffnit, and S. Sinsel, *Hardware-in-the-loop simulation for the design and testing of engine-control systems.* Control Engineering Practice, 1999. **7**: p. 643–653.
4. MATLAB-Help. *Software-in-the-loop simulation.* 2020; Software Help. Available from: https://uk.mathworks.com/help/ecoder/software-in-the-loop-sil-simulation.html
5. Lee, W., M. Yoon, and M. Sunwoo, *A cost- and time-effective hardware-in-the-loop simulation platform for automotive engine control systems.* Automobile Engineering, 2003. **217**: p. 41–52.
6. Moskwa, J.J., *Automotive engine modeling for real-time control*, in *Mechanical Engineering.* 1988, Cambridge, MA, USA: Massachusetts Institute of Technology.
7. Moskwa, J.J. and J.K. Hedrick, *Automotive engine modeling for real time control application*, in *American Control Conference.* 1987. IEEE: Minneapolis, MN, USA.
8. Shamekhi, A.H. and A. Ghaffari, *An improved model for SI engines*, in *ASME 2004 Internal Combustion Engine Division Fall Technical Conference.* 2004: Long Beach, CA, USA. p. 215–222.
9. Arsie, I., C. Pianese, and G. Rizzo. *Enhancement of control oriented engine models using neural network*, in *Theory and Practice of Control and Systems, Proceedings of the 6th IEEE Mediterranean Conference.* 1998: Alghero, Sardinia, Italy. p. 465–471. IEEE.
10. Muller, R. and H.H. Hemberger, *Neural adaptive ignition control*, SAE Technical Paper 981057, SAE International. 1998.
11. Gang, X., J. Zhang, and L. Chen, *On-line prediction of engines based on fast neural network*, SAE Technical Paper 2001-01-0562, SAE International. 2001.
12. He, Y. and C.J. Rutland, *Modeling of a turbocharged DI diesel engine using artificial neural networks*, SAE Technical Paper 2002-01-2772, SAE International. 2002.
13. He, Y. and C.J. Rutland, *Application of artificial neural networks in engine modeling.* International Journal of Engine Research, 2004. **5**(4): p. 281–296.

14. He, Y. and C.J. Rutland, Neural *c*ylinder *m*odel and its *t*ransient *r*esults, SAE Technical Paper 2003-01-3232, SAE International. 2003.
15. Brahma, I., Y. He, and C.J. Rutland, *Improvement of neural network accuracy for engine simulations*, SAE Technical Paper 2003-01-3227, SAE International. 2003.
16. Brusca, S., R. Lanzafame, and M. Messina, A *combustion model* for ICE *by means of neural network*, SAE Technical Paper 2005-05-11, SAE International. 2005.
17. Samadani, E., et al., *A method for precalibration of DI diesel engine emissions and performance using neural network and multi-objective genetic algorithm.* Iranian Journal of Chemistry and Chemical Engineering, 2009. **28**(4): p. 61–70.
18. Shivakumar, P.S.P. and B.R. Shrinivasa Rao, *Artificial neural network based prediction of performance and emission characteristics of a variable compression ratio CI engine using WCO as a biodiesel at different injection timings.* Applied Energy, 2011. **88**(7): p. 2344–2354.
19. Uzun, A., *A parametric study for specific fuel consumption of an intercooled diesel engine using a neural network.* Fuel, 2012. **93**: p. 189–199.
20. Yap, W.K. and V. Karri, *Emissions predictive modelling by investigating various neural network models.* Expert Systems with Applications, 2012. **39**(3): p. 2421–2426.
21. Ismail, H.M., et al., *Artificial neural networks modelling of engine-out responses for a light-duty diesel engine fuelled with biodiesel blends.* Applied Energy, 2012. **92**: p. 769–777.
22. Cay, Y., et al., *Prediction of engine performance for an alternative fuel using artificial neural network.* Applied Thermal Engineering, 2012. **37**: p. 217–225.
23. Janakiraman, V.M., X. Nguyen, and D. Assanis, *Nonlinear identification of a gasoline HCCI engine using neural networks coupled with principal component analysis.* Applied Soft Computing, 2013. **13**(5): p. 2375–2389.
24. Molina, S., et al., *Development of a control-oriented model to optimise fuel consumption and* NO_X *emissions in a DI diesel engine.* Applied Energy, 2014. **119**: p. 405–416.
25. Roy, S., R. Banerjee, and P.K. Bose, *Performance and exhaust emissions prediction of a CRDI assisted single cylinder diesel engine coupled with EGR using artificial neural network.* Applied Energy, 2014. **119**: p. 330–340.
26. Rezaei, J., et al., *Performance prediction of HCCI engines with oxygenated fuels using artificial neural networks.* Applied Energy, 2015. **138**: p. 460–473.
27. Kapusuz, M., H. Ozcan, and J. Ahmad Yamin, *Research of performance on a spark ignition engine fueled by alcohol-gasoline blends using artificial neural networks.* Applied Thermal Engineering, 2015. **91**: p. 525–534.
28. Tosun, E., K. Aydin, and M. Bilgili, *Comparison of linear regression and artificial neural network model of a diesel engine fueled with biodiesel-alcohol mixtures.* Alexandria Engineering Journal, 2016. **55**(4): p. 3081–3089.
29. Acharya, N., et al., *An artificial neural network model for a diesel engine fuelled with mahua biodiesel*, in *Computational Intelligence in Data Mining.* 2017, Berlin, Germany: Springer. p. 193–201.
30. Gürgen, S., B. Ünver, and I. Altın, *Prediction of cyclic variability in a diesel engine fueled with n-butanol and diesel fuel blends using artificial neural network.* Renewable Energy, 2018. **117**: p. 538–544.
31. Yu, M., et al., *Diesel engine modeling based on recurrent neural networks for a hardware-in-the-loop simulation system of diesel generator sets.* Neurocomputing, 2018. **29**: p. 9–19.
32. Di Mauro, A., H. Chen, and V. Sick. Neural network prediction of cycle-to-cycle power variability in a spark-ignited internal combustion engine. Proceedings of the Combustion Institute. 2019. **37**(4): p. 4937–4944.

33. Shamekhi, A.-M. and A.H. Shamekhi, *A new approach in improvement of mean value models for spark ignition engines using neural networks*. Expert Systems with Applications, 2015. **42**: p. 5192–5218.
34. Gerhardt, J., N. Benninger, and W. Hess, *Torque-oriented functional structure of an electronic engine management as a new basis of drivetrain systems (in German)*, in *6th Aachener Kolloquium Fahrzeug-und Motorentechnik*. 1997: Aachen, Germany.
35. Gerhardt, J., H. Hönninger, and H. Bischof, A new approach to functional and software structure for engine management systems - Bosch ME97, Technical Paper 1998-02-23, SAE International. 1998.
36. Shamekhi, A.-M. and A.H. Shamekhi, *Engine model-based pre-calibration and optimization for mid-level hierarchical control design*. SAE International Journal of Engines, 2021. **14**(5): p. 651–669.
37. Rask, E. and M. Sellnau, Simulation-*based engine calibration: tools, techniques, and applications*, Technical Paper 2004-01-1264, SAE International. 2004: Detroit, MI, USA.
38. Zhao, J. and M. Xu, *Fuel economy optimization of an Atkinson cycle engine using genetic algorithm*. Applied Energy, 2013. **105**: p. 335–348.
39. Bellis, V.D., *Performance optimization of a spark-ignition turbocharged VVA engine under knock limited operation*. Applied Energy, 2016. **164**: p. 162–174.
40. Jiang, S., D. Nutter, and A. Gullitti, Implementation of *model-based calibration for a gasoline engine*, Technical Paper 2012-01-0722, SAE International. 2012.
41. Grasreiner, S., et al., *Model-based virtual engine calibration with the help of phenomenological methods for spark-ignited engines*. Applied Thermal Engineering, 2017. **121**: p. 190–199.
42. Langouët, H., et al., *Optimization for Engine Calibration*, in *International Conference on Engineering Optimization*. 2008: Rio de Janeiro, Brazil.
43. Langouët, H., et al., *Engine calibration: multi-objective constrained optimization of engine maps*. Optimization and Engineering, 2011. **12**: p. 407–424.
44. Isermann, R. and H. Sequenz, *Model-based development of combustion-engine control and optimal calibration for driving cycles: general procedure and application*. IFAC-PapersOnLine, 2016. **49**(11): p. 633–640.
45. Park, S., et al., *Optimization and calibration strategy using design of experiment for a diesel engine*. Applied Thermal Engineering, 2017. **123**: p. 917–928.
46. Millo, F., P. Arya, and F. Mallamo, *Optimization of automotive diesel engine calibration using genetic algorithm techniques*. Energy, 2018. **158**: p. 807–819.
47. Hafner, M. and R. Isermann, *Multiobjective optimization of feedforward control maps in engine management systems towards low consumption and low emissions*. Transactions of The Institute of Measurement and Control, 2003. **25**: p. 57–74.
48. Atkinson, C. and G. Mott, Dynamic model-based calibration optimization: An introduction and application to diesel engines, Technical Paper 2005-01-0026, SAE International. 2005: Detroit, MI, USA.
49. Atkinson, C., Fuel efficiency optimization using rapid transient engine calibration, Technical Paper 2014-01-2359, SAE International. 2014.
50. Schaberg, P. and C. Atkinson, Calibration optimization of a heavy-duty diesel engine with GTL diesel fuel, Technical Paper 2016-01-0622, SAE International. 2016.
51. Arya, P., F. Millo, and F. Mallamo, *A fully automated smooth calibration generation methodology for optimization of latest generation of automotive diesel engines*. Energy, 2019. **178**: p. 334–343.
52. Tayarani-N, M.-H., X. Yao, and H. Xu, *Meta-heuristic algorithms in car engine design: A literature survey*. IEEE Transactions on Evolutionary Computation, 2015. **19**(5): p. 609–629.

53. Malikopoulos, A.A., P.Y. Papalambros, and D.N. Assanis. *A Learning Algorithm for Optimal Internal Combustion Engine Calibration in Real Time*, in *Proceedings of the ASME 2007 International Design Engineering Technical Conferences & Computers and Information in Engineering Conference*. 2007. ASME: Las Vegas, NV, USA.
54. Malikopoulos, A.A., D.N. Assanis, and P.Y. Papalambros, *Real-time self-learning optimization of diesel engine calibration*. Journal of Engineering for Gas Turbines and Power, 2009. **131**(2): 022803.
55. Boes, J., et al., Model-*free optimization of an engine control unit thanks to self-adaptive multi-agent sys*tems, in International Conference on Embedded Real Time Software and Systems – ERTS. 2015, HAL: Toulouse, France. p. 350–359.
56. Wong, P.K., et al., *Online extreme learning machine based modeling and optimization for point-by-point engine calibration. Neurocomputing*, 2017. **277**: p. 187–197.
57. Nikzadfar, K. and A.H. Shamekhi, *Investigating a new model-based calibration procedure for optimizing the emissions and performance of a turbocharged diesel engine*. Fuel, 2019. **242**: p. 455–469.
58. Maloney, P.J., Objective *determination of minimum engine mapping requirements for optimal* SI DIVCP *engine calibration*, Technical Paper 2009-01-0246, SAE International. 2009.
59. Shamekhi, A.-M. and A.H. Shamekhi, *Engine adaptive fuel injection control using model predictive self-tuning regulator with adaptive variable functioning*. SAE International Journal of Engines, 2021. **14**(5): p. 671–682.
60. Stotsky, A. and I. Kolmanovsky, *Application of input estimation techniques to charge estimation and control in automotive engines*. Control Engineering Practice, 2002. **10**: p. 1371–1383.
61. Arsie, I., et al., *An adaptive estimator of fuel film dynamics in the intake port of a spark ignition engine*. Control Engineering Practice, 2003. **11**: p. 303–309.
62. Albrecht, A., et al., *Observer design for downsized gasoline engine control using 1D engine simulation*. Oil & Gas Science and Technology – Rev. IFP, 2006. **61**(1): p. 165–179.
63. Le Solliec, G., et al., *Engine control of a downsized spark ignited engine: From simulation to vehicle*. Oil & Gas Science and Technology – Rev. IFP, 2007. **62**(4): p. 555–572.
64. Meng, L., et al., *Intake air mass observer design based on extended Kalman filter for air-fuel ratio control on SI engine*. Energies, 2019. **12**(18): p. 3444.
65. Wua, H.-M. and R. Tafreshi, *Observer-based internal model air–fuel ratio control of lean-burn SI engines*. IFAC Journal of Systems and Control, 2019. **9**: 100065.
66. Tang, H., et al., *Adaptive and learning control for SI engine model with uncertainties*. IEEE/ASME Transactions on Mechatronics, 2009. **14**(1): p. 93–104.
67. Tang, H., et al., *Engine control design using globally linearizing control and sliding mode*. Transactions of the Institute of Measurement and Control, 2010. **32**(2): p. 225–247.
68. Liu, D., et al., *Adaptive critic learning techniques for engine torque and air–fuel ratio control*. IEEE Transactions on Systems, Man, and Cybernetics—Part B: Cybernetics, 2008. **38**(4): p. 988–993.
69. Muske, K.R., J.C. Peyton Jones, and E.M. Franceschi, *Adaptive analytical model-based control for SI engine air–fuel ratio*. IEEE Transactions on Control Systems Technology, 2008. **16**(4): p. 763–768.
70. Ebrahimi, B., et al., *A parameter-varying filtered PID strategy for air–fuel ratio control of spark ignition engines*. Control Engineering Practice, 2012. **20**(8): p. 805–815.
71. Efimov, D.V., V.O. Nikiforov, and H. Javaherian, *Supervisory control of air–fuel ratio in spark ignition engines*. Control Engineering Practice, 2014. **30**: p. 27–33.

72. Zhou, X., et al., *Modeling and control of the air path system in turbocharged gasoline engine*, in *27th Chinese Control and Decision Conference*. 2015: Qingdao, China. p. 3469–3474.
73. Kumar, M. and T. Shen, *In-cylinder pressure-based air-fuel ratio control for lean burn operation mode of SI engines*. Energy, 2017. **120**: p. 106–116.
74. Song, K., T. Hao, and H. Xie, *Disturbance rejection control of air–fuel ratio with transport-delay in engines*. Control Engineering Practice, 2018. **79**: p. 36–49.
75. Na, J., et al., *Air-fuel ratio control of spark ignition engines with unknown system dynamics estimator: Theory and experiments*. IEEE Transactions on Control Systems Technology, 2019. **29**: p. 1–8.
76. Wagner, J.R., D.M. Dawson, and L. Zeyu, *Nonlinear air-to-fuel ratio and engine speed control for hybrid vehicles*. IEEE Transactions on Vehicular Technology, 2003. **52**(1): p. 184–195.
77. Piltan, F., et al., *Adjust the fuel ratio by high impact chattering free sliding methodology with application to automotive engine*. International Journal of Hybrid Information Technology, 2013. **6**(1): p. 13–24.
78. Wu, H.-M. and R. Tafreshi, *Fuzzy sliding-mode strategy for air–fuel ratio control of lean-burn spark ignition engines*. Asian Journal of Control, 2018. **20**(1): p. 149–158.
79. Powell, J.D., *Engine control using cylinder pressure: Past, present, and future*. Journal of Dynamic Systems, Measurement, and Control, 1993. **115**: p. 343–350.
80. Auzins, J., H. Johansson, and J. Nytomt, *Ion-gap sense in misfire detection, knock and engine control*, Technical Paper 950004, SAE International, 1995.
81. Sellnau, M.C., F.A. Matekunas, P.A. Battiston, C.-F. Chang, and D.R. Lancaster, *Cylinder-pressure-based engine control using pressure-ratio-management and low-cost non-intrusive cylinder pressure sensors*, Technical Paper 2000-01-0932, SAE International, 2000.
82. Haskara, I., G. Zhu, and J. Winkelman, IC *engine retard ignition timing limit detection and control using in-cylinder ionization signal*, Technical Paper 2004-01-2977, SAE International. 2004.
83. Huang, L.-W. and Y.-G. Chen, *development of car electronic ignition control system based on curving surface-fitting*, in *IEEE International Conference on Vehicular Electronics and Safety*. 2006: Shanghai, China. p. 214–217.
84. Shamekhi, A.H. and A. Ghaffari, *Fuzzy control of spark advance by ion current sensing*. Journal of Automobile Engineering, Proceedings of the Institution of Mechanical Engineers, Part D, 2007. **221**: p. 335–342.
85. Lezius, U., et al., *Improvements in knock control*, in *Mediterranean Conference on Control & Automation*. 2007: Athens, Greece.
86. Behrendt, S., P. Dünow, and B.P. Lampe. *An application of model predictive control to a gasoline engine*, in *Proceedings of 18th International Conference on Process Control*. 2011: Tatranská Lomnica, Slovakia.
87. Malaczynski, G., G. Roth, and D. Johnson, Ion-*sense-based real-time combustion sensing for closed loop engine control*, Technical Paper 2013-01-0354, SAE International. 2013.
88. Tamaki, S., et al. *On-line feedforward map generation for engine ignition timing control*, in *Proceedings of the 19th World Congress of the International Federation of Automatic Control*. 2014: Cape Town, South Africa.
89. Jamali, P., et al., *Weight optimal proportional-integral-plus control of a gasoline engine model*, in *2015 European Control Conference (ECC)*. 2015: Linz, Austria.
90. Zhang, Y., J. Gao, and T. Shen, *Probabilistic guaranteed gradient learning-based spark advance self-optimizing control for spark-ignited engines*. IEEE Transactions on Neural Networks and Learning Systems, 2017. **29**(10): p. 4683–4693.

91. Maldonado, B.P. and A.G. Stefanopoulou, *Cycle-to-cycle feedback for combustion control of spark advance at the misfire limit.* Engineering for Gas Turbines and Power, 2018. **140**(10): 102812.
92. Lee, J., et al., *Effect of different excess air ratio values and spark advance timing on combustion and emission characteristics of hydrogen-fueled spark ignition engine.* Hydrogen Energy, 2019. **44**: p. 25021–25030.
93. Lee, B., G. Rizzoni, Y. Guezennec, A. Soliman, M. Cavalletti, and J. Waters, *Engine control using torque estimation*, Technical Paper 2001-01-0995, SAE International. 2001.
94. Jaine, T., A. Charlet, P. Higelin, and Y. Chamaillard, High *frequency* IMEP *estimation and filtering for torque based* SI *engine control*, Technical Paper 2002-03-04, SAE International. 2002.
95. Stotsky, A. and A. Forgo, *Recursive spline interpolation method for real time engine control applications.* Control Engineering Practice, 2004. **12**: p. 409–416.
96. Helm, S., M. Kozek, and S. Jakubek, *Combustion torque estimation and misfire detection for calibration of combustion engines by parametric Kalman filtering.* IEEE Transactions on Industrial Electronics, 2011. **59**(11): p. 4326–4337.
97. Falcone, P., G. Fiengo, and L. Glielmo, *Non-linear net engine torque estimator for internal combustion engine.* IFAC Proceedings Volumes, 2004. **37**(22): p. 125–130.
98. Falcone, P., G. Fiengo, and L. Glielmo. Nicely *nonlinear engine torque estimator*, in Proceedings of IFAC World Congress. 2005, Prague, Czech Republic.
99. Butt, Q.R. and A.I. Bhatti, *Estimation of gasoline-engine parameters using higher order sliding mode.* IEEE Transactions on Industrial Electronics, 2008. **55**(11): p. 3891–3898.
100. Ahmed, Q. and A.I. Bhatti, *Estimating SI engine efficiencies and parameters in second-order sliding modes.* IEEE Transactions on Industrial Electronics, 2011. **58**(10): p. 4837–4846.
101. Oh, S., et al., *Real-time IMEP estimation and control using an in-cylinder pressure sensor for a common-rail direct injection diesel engine.* Journal of Engineering for Gas Turbines and Power, 2011. **133**(6): 062801.
102. Na, J., et al., *Vehicle engine torque estimation via unknown input observer and adaptive parameter estimation.* IEEE Transactions on Vehicular Technology, 2018. **67**(1): p. 409–422.
103. Chamaillard, Y., P. Higelin, and A. Charlet, *A simple method for robust control design, application on a non-linear and delayed system: Engine torque control.* Control Engineering Practice, 2004. **12**: p. 417–429.
104. Huang, T., et al. *Neural sliding-mode control of engine torque*, in *Proceedings of the 17th World Congress of the International Federation of Automatic Control.* 2008: Seoul, South Korea.
105. Bao, W. and H.-F. Kong, *Engine torque requirement control in gear shift process of DCT based on generalized predictive control.* China Journal of Highway and Transport, 2017. **30**(10): p. 145–150.
106. Anjum, R., et al., *Dual loop speed tracking control for torque management of gasoline engines*, in *2019 18th European Control Conference (ECC).* 2019, IEEE: Naples, Italy. p. 3084–3089.
107. Shamekhi, A.-M., A. Taghavipour, and A.H. Shamekhi, *Engine idle speed control using nonlinear multiparametric model predictive control.* Optimal Control: Applications and Methods, 2020. **41**(3): p. 960–979.
108. Shamekhi, A.-M. and A.H. Shamekhi, *Modeling and simulation of combustion in SI engines via neural networks and investigation of calibration and data acquisition in the GT-power software.* Modares Mechanical Engineering, 2015. **14**(13): p. 233–244.

2 Control-Oriented Modeling

2.1 INTRODUCTION

This chapter is to present a real-time, highly accurate, complete model, covering real inputs and outputs necessary for a model in a control system design. In addition to being real time, the accuracy of the model will be comparable to a computational fluid dynamics (CFD) model.

Table 2.1 outlines some of the problems and challenges in the publications reviewed.[i] As noted in Chapter 1, designing a control system for an engine does invoke a complete engine model, including all engine features from the throttle body to flywheel. Only in this way, the dynamic (and transient) behavior of an engine could be captured (with factual engine inputs and outputs).

In what follows, after describing the methodology and data acquisition, engine subsystems will be modeled, and in the end, validation and parametric study will be performed.

2.2 MODELING PRACTICE: NEURO-MVM

As explained in Chapter 1, in the late 1990s, neural networks started being employed instead of mean value models in order to improve model precision. Nevertheless, neural networks inherently suffer from two key downsides. For one thing, as the complexity of the system to be modeled increases, the prediction capability of the network decreases; that is, the accuracy of the neural network declines. For another thing, the reliability of a black-box model, in fact, might not be as high as a white-box model. This problem escalates outside of the trained area. As regards dynamic neural networks, in addition to these two problems, and also the drastic increase in the time required for training, the likelihood of being trapped in local minima overly rises. The number of hidden neurons required to model a system rises with system complexity. Apart from increasing the variance error (as the number of training data is assumed to be constant), this leads the performance surface of the network to become more convoluted, and hence, the probability of being stuck in local minima grows [1–4].

Furthermore, training of dynamic neural networks (using, e.g., pseudo-random binary signals, aka, PRBS) becomes more and more difficult, as the number of inputs increases. It shall be noted that the usage of PRBS for an input may sometimes cause the system to become unstable (since the system outputs might fail to remain within stable operating ranges), which, per se, put a strain on the input. As the number of inputs grows, this issue becomes more and more uncontrollable.

DOI: 10.1201/9781003323044-2

TABLE 2.1
Some of the Major Challenges in the Publications Reviewed Regarding Control-Oriented Modeling

Challenges
Modeling accuracy (especially for global models)
Modeling the engine transient behavior
Covering the widest ranges of engine operation
Covering necessary inputs (e.g. ambient conditions, octane number, etc.)
Covering necessary outputs (e.g. knock, emissions, etc.)
Reliability challenge when totally relying on mere black-box models

This is why, the papers reviewed (regarding dynamic neural networks [NNs]) normally either vary one or two inputs or vary inputs within a limited range, while a real engine involves at least six inputs, each of which operates within a wide range. Accordingly, we avoid modeling the whole engine (as an immensely nonlinear and complex system) using one dynamic neural network.

As explained in [5], another approach is to utilize a large number of dynamic neural network local models within the operating space for modeling. In other words, the operating space is broken down into a number of points as representatives, and local dynamic neural networks are trained around each point. More information in this regard could be found in [5]. Although useful, it still suffers from some of the problems mentioned above, especially the curse of dimensionality. That is, since the number of inputs is high, an overwhelming number of dynamic NN local models are required. Consequently, we prefer to adopt another approach. As will be seen, this approach is very useful, particularly, when it comes to calibration.

The key idea of our modeling style (named Neuro-mean value modeling [MVM]) is to combine black-box models and white-box models so that a gray-box extension is obtained; that is, neural networks are incorporated into subsystems of the mean value model (or discrete event model). In this way, an amalgamation of mathematical-physical insight and neural networks is achieved, and the final model (as will be demonstrated) will enjoy the advantages of both physical-mathematical modeling and black-box modeling. In this way, some knowledge regarding the system's physical structure is incorporated in the model, making the resulting gray-box structure more reliable than a mere black-box model. In other words, this gray-box model will be much more accurate than prevalent mean value models and also more reliable than a mere black-box model in a wide range of operations (as a result of the physical insight incorporated into the model).

Additionally, this will simplify the tasks burdened on neural networks, and as will be shown, the accuracy of the networks will increase; consequently, the accuracy of the entire model will improve. Generally, simplifying each neural network's task, in a sense, ties in with Parsimony Principle and is one of the key points in this study. On top of these, this model provides access to the intermediate parameters of an engine (since all the subsystems are separately identified), which is of paramount significance for a practical simulator. It is illustrated in Section 2.4.1 that gray-box combination could make the model more reliable than a single neural network.

Therefore, the model structure will be comprised of subsystems indicated in Figure 1.6 (re-displayed in Figure 2.1), each one of which contains neural networks.

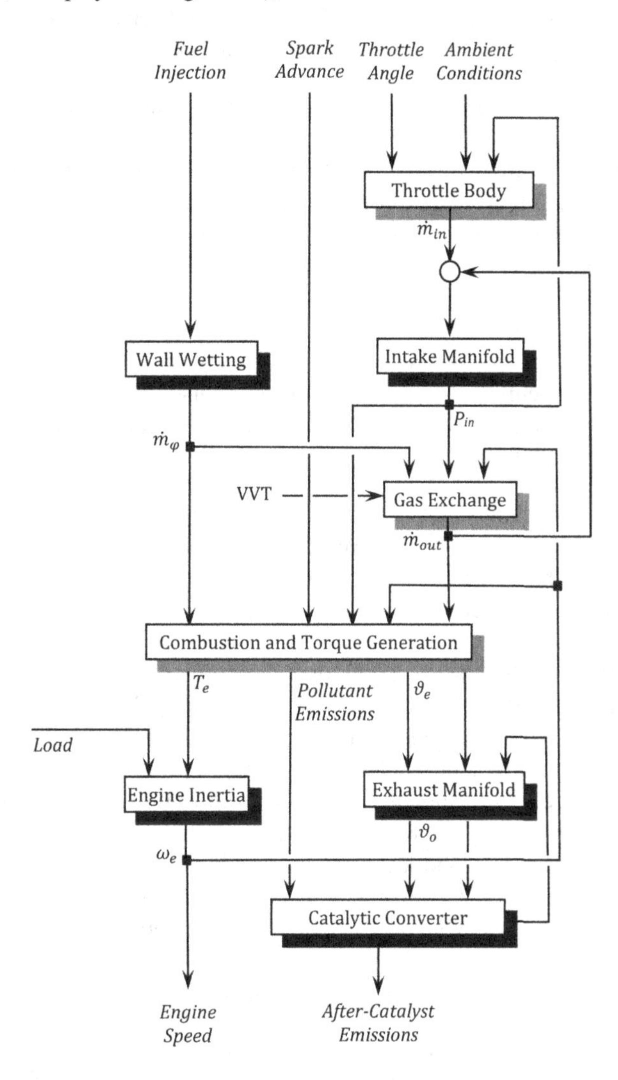

FIGURE 2.1 Re-display of Figure 1.6. SI engine subsystems with their inputs and outputs in MVM (based on [6]). Note that turbocharger and engine temperature subsystems are not shown. Besides, exhaust manifold subsystem could have a different arrangement. Moreover, recall that subsystems, based on their time constant, are divided into three categories: static (with very small time response), constant (very large time response), and dynamic (with medium time response Section 1.3). The blocks with black shadings are dynamic, while the blocks with gray shadings are static.

It is to be noted that, for this problem, shallow neural networks ought to be employed. In other words, application of deep neural networks for this problem would be just like eating soup with a fork! The reason for this is the fact that deep neural networks (e.g. convolutional neural networks [CNNs]) are normally used when a problem involves (filtering) correlation, redundancy, and/or disturbance, none of which exist here. Accordingly, the application of shallow neural networks does suffice.

Methods applied in designing the shallow neural networks are as follows:

- Iteration with different initial weights to make sure of convergence to global minimum [1].
- Utilizing a comprehensive pattern table containing various operating points, and appropriate training, cross-validating, and testing sets [2, 7].
- Adjusting early stopping and avoiding over-fitting [2].
- Paying attention to bias/variance dilemma and avoiding over-determination [2, 8].
- Parsimony principle [8].
- Avoiding irrelevant or less important inputs for neural networks [9].
- Employing neural networks with single outputs [10].
- Using two hidden layers to make the network deeper with higher identification capability.

As the second hidden layer (utilized for deeper learning) combines the results of the first hidden layer for a higher-level classification, it normally contains fewer neurons than the first layer.

Using single output neural networks simplifies the task burdened on networks and enhances the accuracy.

A prominent technique to avoid over-determined networks is to employ "Bayesian Regularization" training algorithm. Using Bayesian framework of David MacKay (implemented within Levenberg–Marquardt algorithm), this algorithm automatically adjusts parameters and prunes redundant ones such that no over-fitting occurs in validation data[ii] [3]. In this study, the first hidden layer would contain around 10 (i.e. 8–12, depending on the complexity of the subsystem) neurons, and the second layer would include about 5 neurons (i.e. 4–8) as the initial architecture. Afterward, Bayesian regularization algorithm would automatically adjust and prune the parameters to find the optimal structure.

On top of the above methods, to achieve higher accuracy (particularly, as to subsystems of higher complexity), the subsequent approaches are employed:

- Committee method: Ensemble averaging [2].
- Improved partitioning method [4], or other local-global models, ANFIS or local model tree models [8].
- Adding assisting knowledge signals as network inputs.

Ensemble averaging (the most famous approach of committee methods) is one of the most effective methods to surmount bias/variance dilemma, notably improving identification quality. It is mathematically proven that an ensemble of slightly over-determined neural networks (i.e. with lower bias error) presents lower variance error (Figure 2.2) [2]. For this purpose, the initial number of hidden neurons could be slightly increased with postponed early stopping.

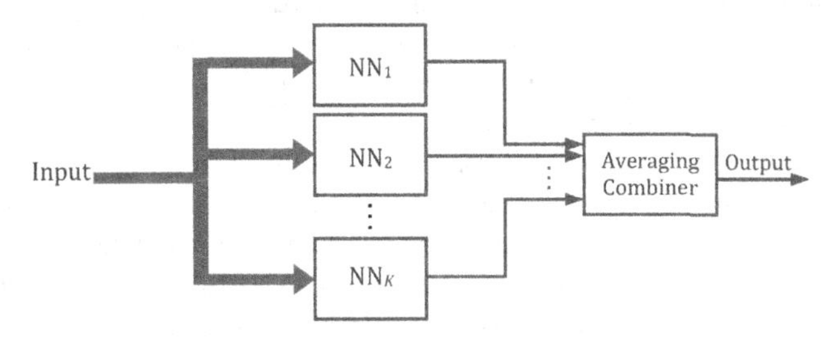

FIGURE 2.2 Ensemble averaging [2].

In the improved partitioning method, to simplify the tasks burdened on neural networks, and thus, to reach more accuracy, the operating space of a subsystem is divided into different sub-spaces (each of which may stand for one operating regime of the subsystem). *Each sub-space shares a part of its volume with the neighboring sub-spaces.* We call this method improved partitioning[iii] [4]. As for each sub-space, a committee of networks can be trained (Figure 2.3). This

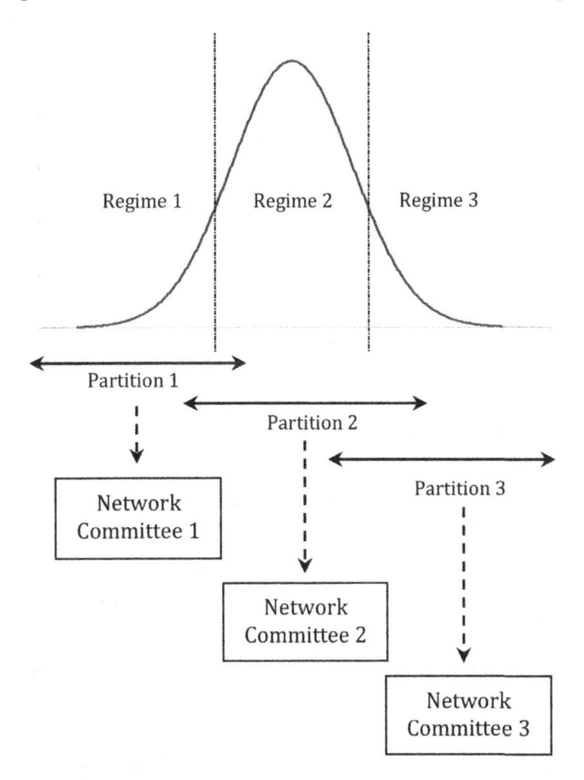

FIGURE 2.3 The combination of ensemble averaging and improved partitioning method. Regime 2 may be divided in two.

combination is significantly effective in improving the accuracy of a model.[iv] The shared spaces compensate for the data reduction for each NN caused by partitioning and also improve the estimation quality for borders.

Adding assisting knowledge signals as network inputs could simplify, and thus, improve identification. However, care needs to be taken so as not to over-determine nor complicate the networks.

In this study, MLP neural networks are employed, as they are stronger in extrapolation compared to RBF networks [8]. On top of that, they give more flexibility in performing the committee method and improved partitioning compared to adaptive neuro-fuzzy inference system (ANFIS) or local linear model tree (LOLIMOT). A short review of the neural networks' design is provided in Appendix A.

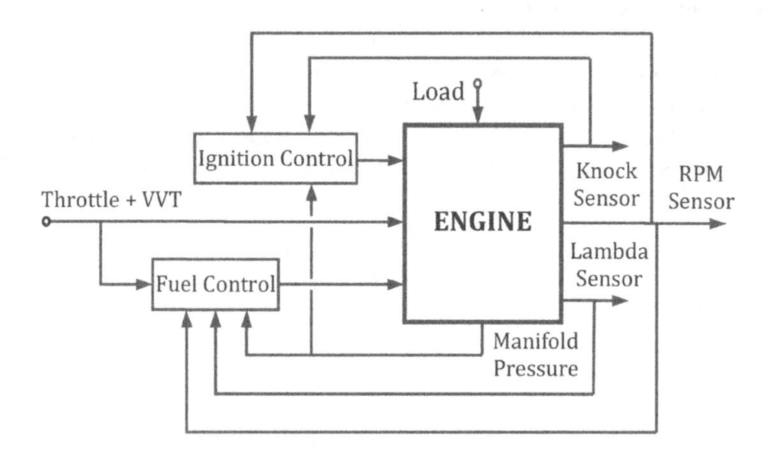

FIGURE 2.4 A simplified relationship between the engine and the engine control unit (inspired from [6]). Ambient conditions are not shown.

As indicated in Figure 2.4, inputs of a real engine *control system* are in fact outputs of the real *engine sensors*. Consequently, the *model* outputs shall include the parameters measured by the sensors in a real engine, namely:

- Engine speed
- Knock
- Lambda
- Manifold pressure

in addition to some other outputs necessary for design, optimization, or parametric study:

- Pollutant emissions
- Produced torque
- Fuel consumption (if necessary)
- Exhaust gas temperature

Model inputs need to be the same as the real engine actuator inputs (that is, controller outputs), along with ambient conditions and also external load:

- Ambient air pressure
- Ambient air temperature
- Throttle angle
- Spark advance
- Injected fuel mass
- Variable valve timing (VVT)
- External load
- Etc. (depending on the engine)

2.2.1 Modeling in Terms of Optimization Perspective

As previously explained, modeling has a profound impact on the calibration quality, even more than the optimization algorithm. As regards optimization (Chapter 3), the model needs to be accurate and real time. More importantly, inasmuch as conventional global models (like MVM) lack accuracy, local models are preferred in the literature; based on which the subsequent optimization maps, however, are likely to require harsh smoothening. This, per se, degrades the accuracy of calibration maps. Therefore, in this study, a *global* transient model with high accuracy and fast response is employed.

Instead of designing a collection of local models, each representing the whole engine for a limited operating interval, each engine subsystem is identified separately for the widest range of engine operation, to make a global model.

As previously explained, this part-to-whole approach coincides with the system's physical structure, as the working fluid flows from one subsystem to another in series. Additionally, it brings several advantages for engine global modeling. That is, the tasks burdened on identifiers are simplified (as opposed to modeling the whole engine using one identifier), and thus, higher levels of accuracy can be achieved. Furthermore, in this way, some knowledge regarding the system's physical structure is incorporated into the model, making the resulting gray-box structure more reliable than a mere black-box model. What is more, this approach provides access to intermediate variables and brings flexibility in optimization.

On top of these, the model includes all the required inputs and outputs for proper optimization.

2.2.2 Assumptions

Modeling assumptions are as follows:

- Along with the experimental data available (which as mentioned are not enough), the previously mentioned validated GT-Power model will be used as the reference system for data acquisition, subsystem identification, and final validation. Obviously, should the real engine with precise testing equipment be available, the reference system would be the real engine, then.
- Wall-wetting subsystem will not be modeled here due to not having access to the exact experimental data required.[v] However, wall-wetting in general will be addressed in Chapter 4.

- Actuators and sensors mostly are not modeled here. Yet, the delay of some sensors has been considered. Should accurate models of actuators and sensors be available, they could also be incorporated to upgrade the model.
- Data acquired from the GT-Power reference model (required due to lack of experimental data), for regions close to idle, and particularly cold start might not be so accurate. More accurate models could be achieved using real experimental data. Yet, the methodology is the same.
- Some special effects (such as scavenging, oil blowby, etc.) might not have been considered in the GT-Power reference model.

Besides, as noted in Section 1.5, the engine in hand for the case study does not contain a turbocharger and EGR valve. These two make small changes in the modeling process, yet, the general methodology is still the same. Some information about their modeling could be found in [6] and one of our previous works [11].

In the end, it needs to be reiterated that there is a facility in GT-Power software that for a designed (1-D) CFD model, conducts data acquisition and then replaces the combustion chamber with a neural network. This facility is utilized by some companies and could be employed by interested readers. However, we prefer to take the approach explained here, for the following reasons:

- Here, neural networks are incorporated into almost all the different subsystems, not merely the combustion subsystem (as it is in GT-Power software).
- Note that GT-Power's control-oriented modeling is completely based on the data acquired from the GT-Power CFD model, itself. Yet, the modeling style proposed here (if the required experimental facilities are available) could be totally based on real experimental data, making it more reliable the software model.
- What is more, the neural network approaches adopted here are much more advanced than a single neural network application utilized in the GT-Power software.

In what follows, after a short glance at data acquisition, modeling the subsystems will be examined.

2.3 DATA ACQUISITION AND DESIGN OF EXPERIMENTS

Even though demanding and time-consuming, this step has a significant impact on system modeling. As explained in Chapter 1, control-oriented modeling is based on experimental data acquired from a real engine test. For the reasons explained in Chapter 1, it *might* be necessary (as it is here) to combine these data with data acquired from a CFD model accurately designed, calibrated, and validated based on the experimental data.

Hence, the experimental data are augmented with data from an accurately validated CFD model (e.g. GT-Power software model), to attain a comprehensive pattern table (covering the whole engine operating space) to train neural networks. More information can be found in our previous paper [12]. We call this combined data "Software-Experimental data." Needless to say, should all the necessary experimental data be available from a precise, reliable engine testing facility, no software data acquisition is required.

In the modeling practice adopted in this study, each engine subsystem can be identified separately. An advantage of Neuro-MVM is the fact that the subsystems to be modeled using neural networks are mostly static, typically demanding steady-state results for identification. As for each subsystem, relevant engine inputs are varied, and the corresponding software-experimental results are acquired for the subsystem inputs/outputs identification.

For appropriate system identification, the whole system operating region shall be covered. Needless to say, densely filling the whole operating space, while greatly improving the identification accuracy, overwhelmingly raises the number of experiments demanded since the number of system inputs is high. Therefore, one of the different "Design of Experiments" approaches (e.g. D-Optimal) could be adopted to enhance the process. Here, we suggest the following strategies developed specifically for this case. Similar to the famous design of experiment methods, this approach is to reduce the number of data required and also their correlations. Yet, unlike other methods that are *general*, these are specifically designed for this very system, and thus, could be more suitable.[vi]

- As stated, each subsystem can be identified individually; thus, of all inputs with the same impact on one specific subsystem, only one (as the relevant variable) needs to be varied, reducing the number of experiments required. For instance, when considering the mass flow rate into the combustion subsystem, amongst throttle angle, and ambient air pressure inputs, only one (e.g. throttle angle, as the most effective one) needs varying; the other can be kept in nominal value.
- Inputs concerning downstream of a subsystem, do not usually have any impact on the subsystem.
- The operating space for inputs influencing a subsystem needs to be partitioned. Within each partition, a value(s) is randomly chosen for the relevant input(s), and the corresponding results are acquired.
- The inputs are divided into three hierarchical categories. The first (highest) category concerns inputs of greatest impacts, influencing almost all the subsystems, namely, engine speed and throttle angle; thus, a number of, e.g., 119 different cases[vii] are regarded for all of the relevant subsystems. The second category (i.e. secondary inputs) pertains to inputs specifically influencing one subsystem, e.g. VVT for the gas-exchange subsystem. As for the respective subsystem, these inputs are varied for each previous (119) case, adding them up to, for instance, 3000 different cases. As can be seen in Figure 2.5, for each (of the primary 119)

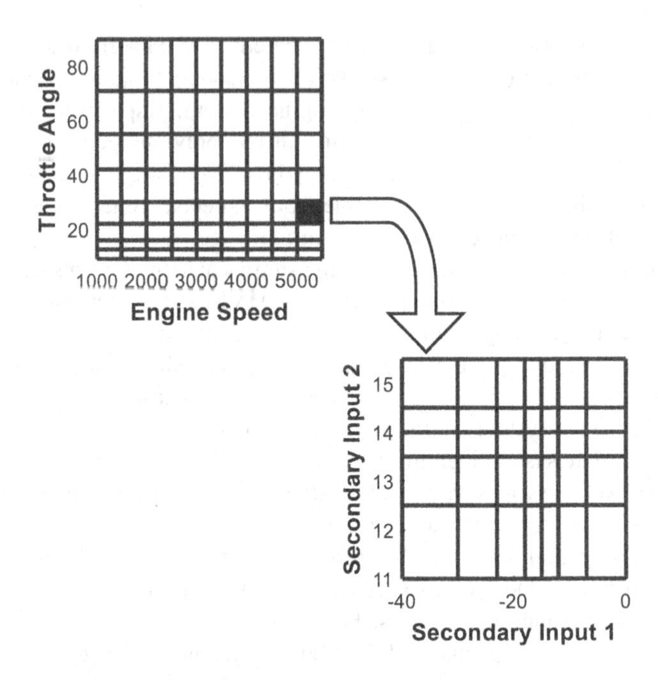

FIGURE 2.5 A simple schematic of partitioning secondary inputs for each primary case. Random values are chosen within each partition. As for throttle angles below 12, meshing is in fact finer than represented.

cases, the relevant secondary inputs are partitioned within their operating space. Each steady-state simulation (for data acquisition) is carried out with inputs randomly chosen within each partition. The pattern for partitioning will be described in the bullet point below.

The lowest category in the hierarchy concerns inputs of the lowest impact on a subsystem, e.g. ambient temperature for the combustion subsystem. These inputs demand larger (looser) partitions or can be randomly allocated for the two aforementioned varying input types (without increasing the number of total cases).

- The inputs are partitioned based on their relevant output behaviors. To optimally minimize data acquisition, the density of partitions can be decreased (increased) as the variation rate of a specific subsystem output goes lower (higher),[viii] as illustrated in Figure 2.6 for VVT.[ix] Physical insight, along with a simple approximate pre-optimization, or a simple surf-fitting through an already known electronic control unit (ECU) point (if available), can be of great help in this regard. As for secondary inputs, there normally exist single optimal points (as in a bell-shaped curve). After approximately locating the optimal point (using a simple approximate pre-optimization or a simple surf-fitting through some already known ECU points), partitioning can be denser (looser) for regions closer (further) to the optimal point, as illustrated in Figure 2.5.

The reason for this is the fact that the region close to the optimal point is more important with faster slope variations[x] (as in a bell-shaped curve). It shall be reiterated that random values are chosen within each partition.

The effects of different ambient air conditions are taken into account for the entire system. Although making data acquisition far more demanding, this results in a practical generalized model.

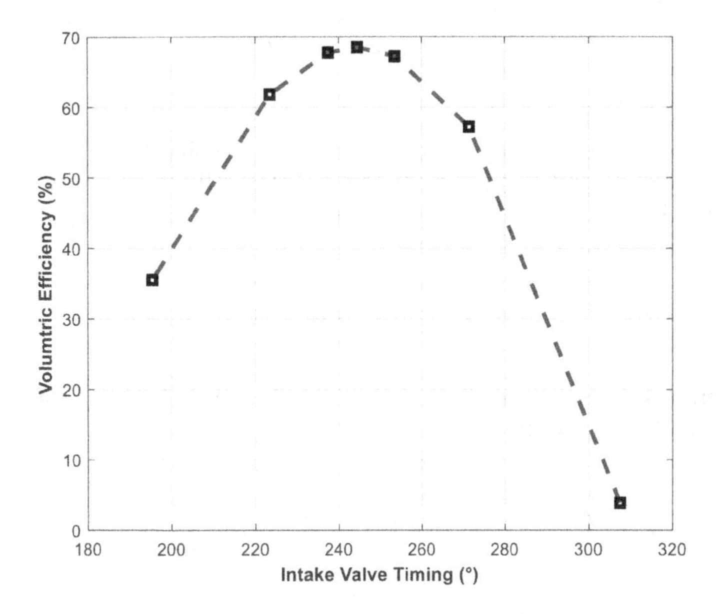

FIGURE 2.6 Changing data acquisition density with output variation rate. As can be seen, the chosen data in the middle are denser.

As a matter of great importance, after a simple pre-optimization for approximately allocating the optimal points, this approach can be readily conducted by simple coding. In what follows, modeling different subsystems is examined in brief.

2.4 MODELING SUBSYSTEMS

2.4.1 Throttle Subsystem

Air flows into the engine through the throttle body. This subsystem is modeled statically. The subsystem inputs are ambient pressure and temperature, throttle angle, and the downstream manifold pressure. The output is the throttle mass flow rate (and temperature). Supposing isentropic expansion, in mean value and discrete event modeling, intake air mass flow can be expressed as follows [6, 13]:

$$\dot{m}(t) = C_d \cdot A(t) \cdot \frac{P_{\text{in}}(t)}{\sqrt{R.\vartheta_{\text{in}}(t)}} \cdot \Psi\left(\frac{P_{\text{in}}(t)}{P_{\text{out}}(t)}\right), \tag{2.1}$$

where $\Psi(\cdot)$ is defined as follows:

$$\Psi\left(\frac{P_{in}(t)}{P_{out}(t)}\right)=\begin{cases}\sqrt{k\left[\frac{2}{k+1}\right]^{\frac{k+1}{k-1}}}, & P_{out}<P_{cr}\\ \left[\frac{P_{out}}{P_{in}}\right]^{1/k}\cdot\sqrt{\frac{2k}{k-1}\cdot\left[1-\left(\frac{P_{out}}{P_{in}}\right)^{\frac{k-1}{k}}\right]}, & P_{out}>P_{cr}\end{cases}. \tag{2.2}$$

In above relations, P_{in} and ϑ_{in}, P_{out}, and k stand for ambient pressure and temperature, intake manifold pressure, and specific heat ratio of air, respectively. P_{cr}, which is critical pressure, is defined as follows:

$$P_{cr}=\left[\frac{2}{k+1}\right]^{\frac{k}{k-1}}\cdot P_{in}. \tag{2.3}$$

"A" represents the effective intake area of throttle body, whose relation with the throttle angle can be expressed as [13]

$$\frac{4\,A}{\pi\,D^2}=\left(1-\frac{\cos\theta}{\cos\theta_0}\right)+\frac{2}{\pi}\left\{\frac{a}{\cos\theta}\left(\cos^2\theta-a^2\cos^2\theta_0\right)^{1/2}\right.$$
$$\left.-\frac{\cos\theta}{\cos\theta_0}\arcsin\left[\frac{a\cos\theta}{\cos\theta_0}\right]-a\left[1-a^2\right]^{\frac{1}{2}}+\arcsin a\right\}. \tag{2.4}$$

In the above relation, D represents the inside diameter of throttle, a denotes the ratio of plate rod diameter to the plate diameter, θ_0 is the angle of throttle plate when closed, and θ represents the throttle angle (from the closed position).

C_d, discharge coefficient, is an experimental parameter, differing for various operating points. In accordance with our gray-box modeling strategy, this part can be modeled with (a committee of) neural networks, yet, based on all the four subsystem inputs. In other words, the network structure, in a sense, modifies the deviation of the mathematical relations from the software-experimental data. For a dense and extensive table of patterns (as in this study), the entire subsystem had better be modeled with a committee of networks (Figure 2.7).

The temperature change through throttle may be tangible. In this study, the impact of ambient temperature is taken into consideration for the whole system. Accordingly, a parallel network is designed to predict the temperature change. Figure 2.8 displays the overall structure of the mean-value-based model for the subsystem.

A table of around 1100 patterns is utilized for training, validation, and test. The networks' initial two-hidden-layer architecture for training contain 8–4 neurons

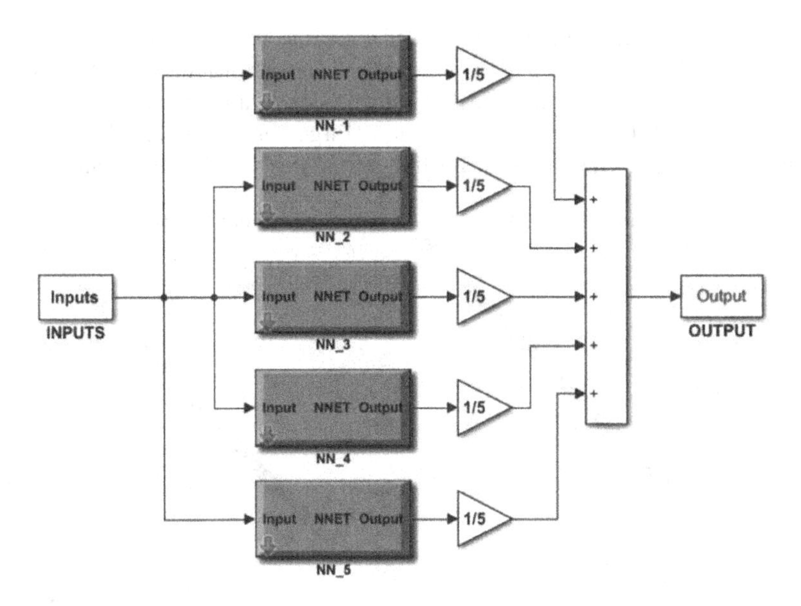

FIGURE 2.7 The committee structure for air mass flow rate through the throttle.

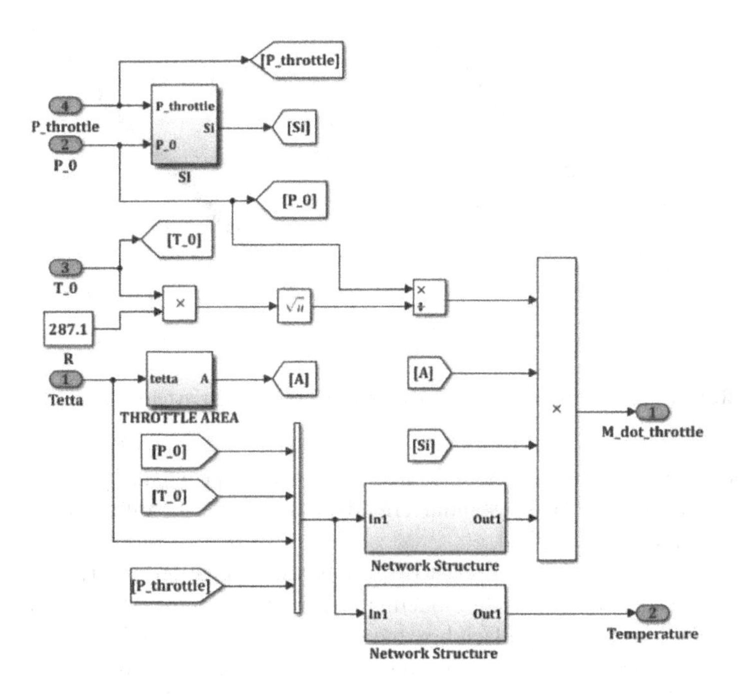

FIGURE 2.8 The overall structure of throttle subsystem in mean-value-based modeling. For a dense and extensive table of patterns (as in this study), the entire subsystem is modeled with two committees of networks representing mass flow rate and temperature.

(i.e. 8 neurons in the first hidden layer, and 4 neurons in the second hidden layer). As explained before, "Bayesian Regularization" algorithm automatically finds the optimal structure. Figure 2.9 depicts the regression of a network applied for the subsystem. Henceforth, in regression plots, "Target" is the target value *expected* from an identifier. "Output" is the resulting *estimated* value by the identifier.

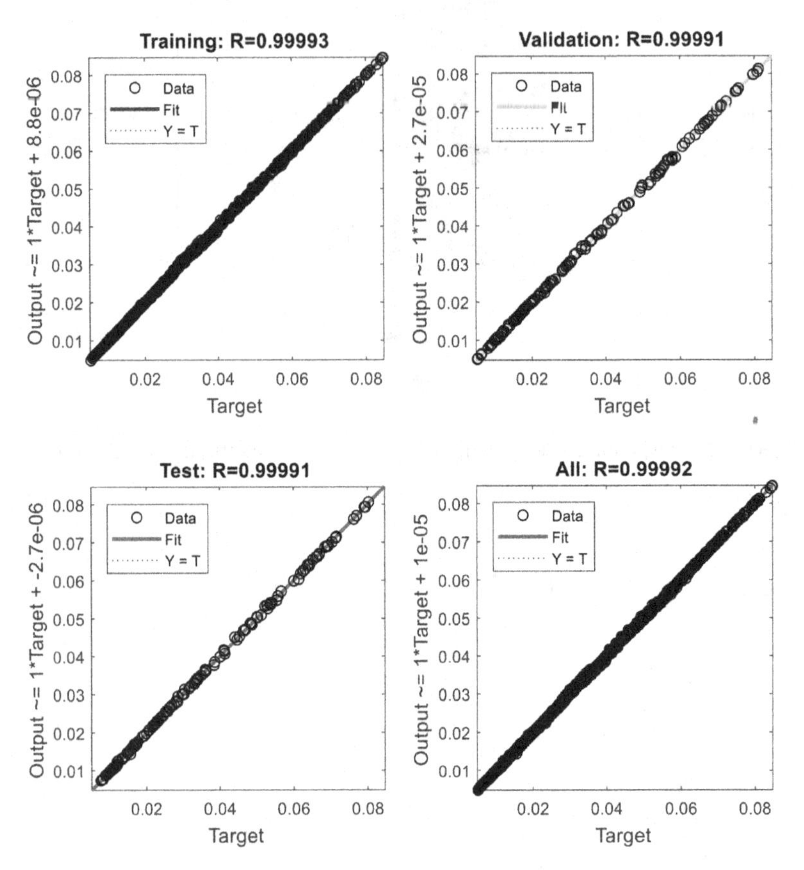

FIGURE 2.9 The regression of a network applied for the subsystem.

As can be seen, the regression is line-like with uniform distribution of points, which is expressive of substantial effectiveness of the identification and data acquisition methods applied.

Now, to investigate the impact of gray-box combination on prediction reliability, consider throttle body modeling structure in Figure 2.8. This structure is compared with a fully black-box model containing only one neural network representing the whole subsystem. Now both are challenged with data outside of the trained region. As can be seen in Figures 2.10 and 2.11, the gray-box structure can come off better (by demonstrating a better regression line) and is more reliable than the fully black-box model.

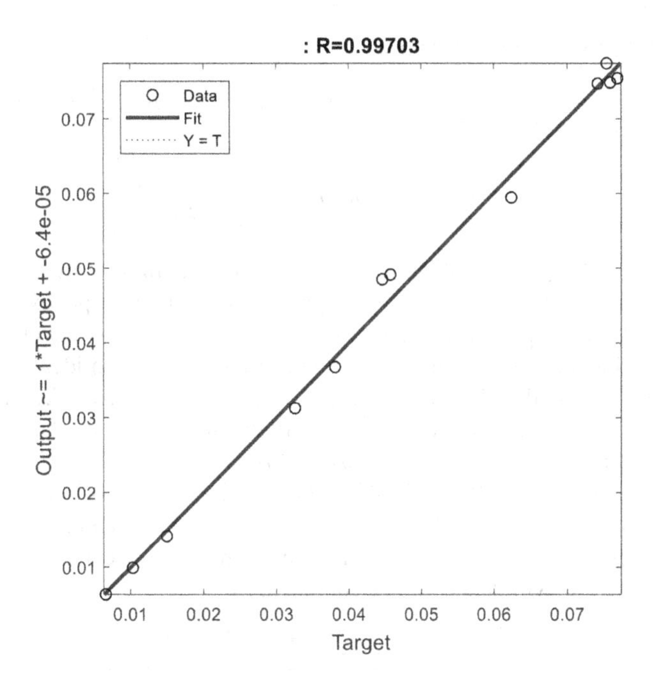

FIGURE 2.10 The regression of the subsystem modeled by gray-box structure for data outside of the trained region.

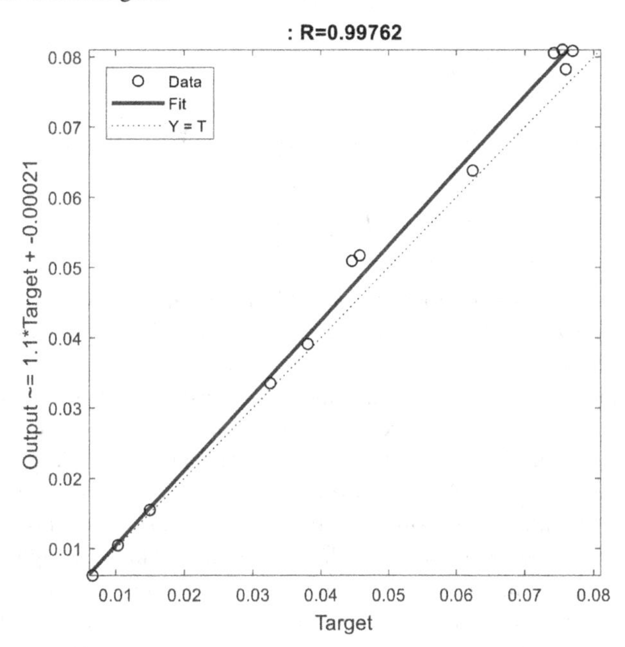

FIGURE 2.11 The regression of the subsystem modeled by one black-box neural network for data outside of the trained region. As can be seen, the regression is not as good as that of the gray-box structure.

2.4.2 Gas Exchange Subsystem

Engine can be assumed as a volumetric pump, whose mass flow rate into the *cylinder* is estimated as [6]

$$\dot{m}(t) = \rho_{\text{in}}(t) \cdot \lambda_l (P_m, \omega_e) \cdot \frac{V_d \cdot \omega_e(t)}{4\pi}, \tag{2.5}$$

where ρ_{in}, $\omega_e(t)$, P_m, and V_d stand for density of the fluid into the cylinder, engine speed (rad/s), manifold pressure, and displacement volume, respectively. As can be seen, this subsystem is also static and models the mass flow rate out of manifold into cylinders. The deviation of the above model from an ideal volumetric pump is compensated with an experimental volumetric coefficient λ_l, making it a great candidate for network incorporation into the mathematical relation.

For port fuel injection SI engines (as in this study) a mixture of air and fuel enters the cylinder. Hence, the density of the mixture, ρ_{in}, needs to be considered in the equation. Assuming complete evaporation of fuel, with adiabatic blending of fuel and air, and also regarding the final mixture as a perfect gas, the amount of air entering the cylinders could be computed as

$$\dot{m}_\beta = \frac{-b + \sqrt{b^2 - 4a \cdot c}}{2a}, \tag{2.6}$$

where

$$\begin{aligned} a &= R_\beta \vartheta_\beta c_{P,\beta} \\ b &= \left(R_\beta \vartheta_\varphi c_{P,\varphi} + R_\varphi \vartheta_\beta c_{P,\beta}\right)\dot{m}_\varphi - P_m V_d \lambda_l \frac{\omega_e}{4\pi} c_{P,\beta} \\ c &= \left(\dot{m}_\varphi R_\varphi \vartheta_\varphi - P_m V_d \lambda_l \frac{\omega_e}{4\pi}\right) c_{P,\varphi} \dot{m}_\varphi. \end{aligned} \tag{2.7}$$

In the above relations, the subscripts φ and β represent fuel and air, respectively. $c_{P,*}$ denotes specific heat at constant pressure, ϑ stands for temperature, and R_* is the respective gas constant.

As mentioned earlier, the experimental volumetric coefficient can be modeled with a committee of networks receiving all the inputs (as we did in our paper [4]). However, as the pattern table available here is extensive (containing 2163 sets of data) and to avoid the possibility of algebraic loop,[xi] the entire subsystem is modeled with a committee of networks (Figure 2.12). As the subsystem is not so complicated, a committee of, for example, five networks would be enough. Too many networks in a committee structure might increase the computational time and shall only be employed if fewer networks fail to bring the accuracy demanded.

The subsystem inputs are: 1) engine speed, 2) intake manifold pressure,[xii] 3) manifold temperature, 4) fuel mass flow rate, 5) fuel temperature, and of course, 6) VVT. The output is air mass flow rate into the cylinder. The biggest challenge in data acquisition for this part pertains to VVT. Since the model will

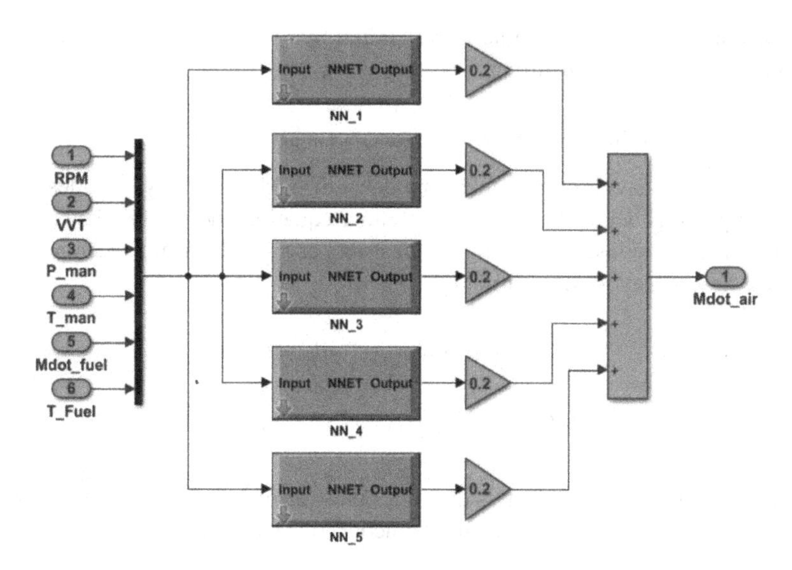

FIGURE 2.12 Ensemble averaging for a committee of five networks representing gas exchange subsystem.

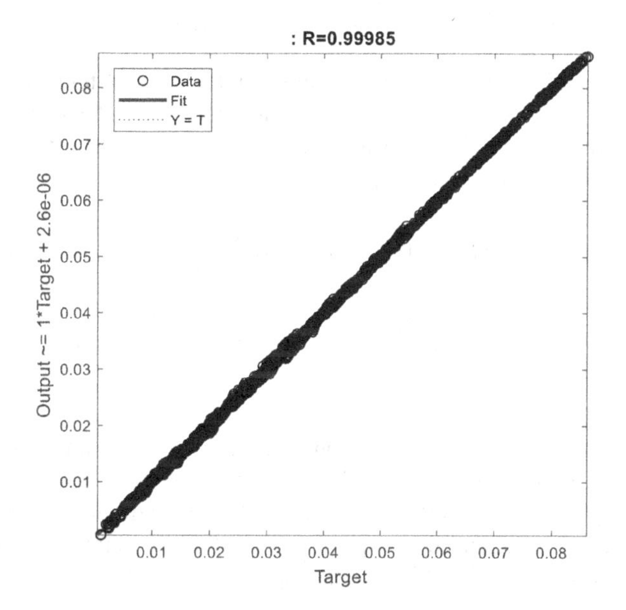

FIGURE 2.13 The overall regression of the gas exchange subsystem.

be used for optimization, it would be of great importance for the model to accurately predict optimum VVTs. This requires enhanced data acquisition. More information in this regard can be found in Section 2.3 and Figure 2.6. The overall regression (including training, validation, and test data) of the subsystem is indicated in Figure 2.13, which is again expressive of substantial accuracy in the subsystem identification. The networks' initial two-hidden-layer architecture for training contained 8–4 neurons.

2.4.3 Receiver Subsystem (Intake Manifold Dynamic)

The dynamic behavior of manifold is modeled through this subsystem, described by the equations below. In fact, the continuity equation and the first law of thermodynamic (i.e. conservation laws of mass and energy) determine accumulation of mass and energy in the manifold:

$$\frac{d}{dt}m(t) = \dot{m}_{\text{in}}(t) - \dot{m}_{\text{out}}(t), \tag{2.8}$$

$$\frac{d}{dt}U(t) = \dot{H}_{\text{in}}(t) - \dot{H}_{\text{out}}(t) + \dot{Q}(t), \tag{2.9}$$

where $\dot{m}_{\text{in}}$ and $\dot{m}_{\text{out}}$ represent mass flow rates into and out of manifold. The internal energy U and flow enthalpies $\dot{H}_*$ are defined as (for ideal gas):

$$\begin{aligned} U(t) &= c_v \cdot \vartheta(t) \cdot m(t), \\ \dot{H}_{\text{in}}(t) &= c_p \cdot \vartheta_{\text{in}}(t) \cdot \dot{m}_{\text{in}}(t), \\ \dot{H}_{\text{out}}(t) &= c_p \cdot \vartheta(t) \cdot \dot{m}_{\text{out}}(t), \end{aligned} \tag{2.10}$$

where and ϑ_{in} and ϑ_{out} denote the upstream and downstream temperatures. Note that in this lumped parameter model, ϑ_{out} is assumed to be the same as the subsystem temperature $\vartheta(t)$. Using ideal gas law and after substitution, the following coupled equations will be derived [6]:

$$\begin{aligned} \frac{d}{dt}P(t) &= \frac{R \cdot k}{V}\left[\dot{m}_{\text{in}}(t) \cdot \vartheta_{\text{in}}(t) - \dot{m}_{\text{out}}(t) \cdot \vartheta(t)\right] \\ \frac{d}{dt}\vartheta(t) &= \frac{R \cdot \vartheta}{P \cdot V \cdot c_v}\left[c_p \cdot \dot{m}_{\text{in}} \cdot \vartheta_{\text{in}} - c_p \cdot \dot{m}_{\text{out}} \cdot \vartheta - c_v \cdot \left(\dot{m}_{\text{in}}(t) - \dot{m}_{\text{out}}(t)\right) \cdot \vartheta(t)\right], \end{aligned} \tag{2.11}$$

where V is manifold volume, and P and ϑ are manifold instantaneous pressure and temperature, respectively. Also, k is the ratio of specific heats c_p to c_v. The subsystem inputs are the throttle mass flow rate and temperature, and the air mass flow into the cylinder; and the outputs are manifold pressure and temperature. Figure 2.14 portrays the Simulink block diagram of the subsystem.

2.4.4 Rotational Dynamics

Flywheel and rotational parts act like a reservoir storing the rotational energy caused from the difference of the engine generated torque and external load [6]:

$$\Theta_e \frac{d}{dt}\omega_e(t) = T_e(t) - T_l(t). \tag{2.12}$$

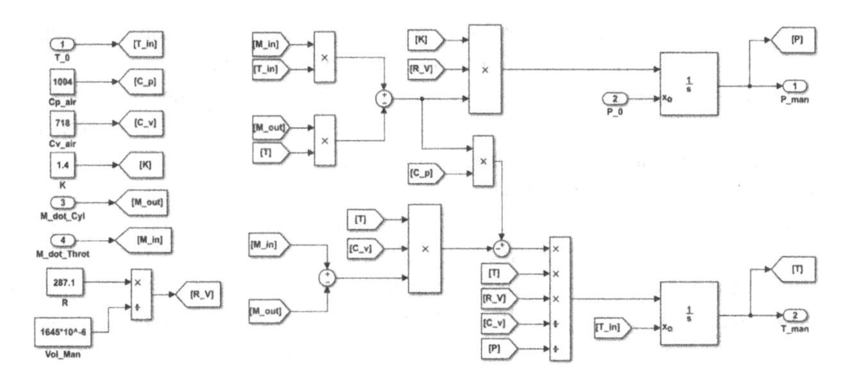

FIGURE 2.14 The Simulink block diagram of the receiver subsystem.

In the above relation, Θ_e, T_e, and T_l represent rotational inertia, generated brake torque, and external load, respectively. The output of the subsystem is engine speed, ω_e.

2.4.5 Combustion Subsystem

One of the most challenging subsystems significantly influencing the whole engine performance is the combustion subsystem. As the valves close in every engine cycle, the boundary conditions during combustion remain constant. Accordingly, the combustion subsystem can be assumed static [6]. The subsystem outputs are as follows:

- Generated brake torque
- Knock detection (knock probability and knock index)
- Exhaust gas temperature
- Emissions:
  - NOx
  - CO
  - UHC (aka. HC)

Emissions are modeled in both ppm (parts per million), and g/kW-h (or BS, i.e., brake specific). The inputs are as follows:

- Engine speed
- Spark advance
- Manifold pressure and temperature
- Air mass flow rates into the cylinders
- Fuel mass flow rates into the cylinders
- Fuel octane number, which we add for knock detection

For the sake of brevity, the engine power and fuel consumption are not shown, as they can be easily obtained by multiplying torque and speed, and dividing the fuel injected by power, respectively. What is more, manifold pressure and temperature are added as combustion inputs to improve the model accuracy.

Appropriate data acquisition plays an essential role for proper identification of this subsystem, even more than previous ones. In this way, the approaches explained in Section 2.3 are suggested.

It is to be noted that in MVM, the discrete reciprocal behavior of engine is captured by introducing time delays. For instance, a change in manifold pressure does not immediately affect engine torque. There is a time delay of τ_{IPS} for engine torque to respond to manifold pressure variation, which represents the induction-to-power-stroke (IPS, almost half a cycle in a four-stroke SI engine) delay:

$$\tau_{IPS} \approx \frac{2\pi}{\omega_e}. \tag{2.13}$$

Also, any change in the mixture entering the cylinder is sensed in the exhaust after a time delay of τ_{IEG}, which is induction-to-exhaust-gas (IEG, almost three-fourth of a cycle) delay [6]:

$$\tau_{IEG} \approx \frac{3\pi}{\omega_e}. \tag{2.14}$$

To simplify the tasks burdened on identifiers, and thus, to improve accuracy, each output is modeled as a separate subsystem as follows. Ensemble averaging of network committees is employed for variance error reduction and better accuracy. For incomprehensive tables of patterns, a combination of committee methods and improved partitioning method (explained in Section 2.2) or local-global modeling could be utilized [4].

The overall structure of the combustion subsystem can be seen in Figure 2.15. In the end, it shall be noted that it is possible to merge this subsystem with "Gas Exchange" subsystem.

2.4.5.1 Torque Generation Subsystem

This most influential subsystem is modeled using ensemble averaging of networks. Thanks to appropriate data acquisition, all the two-hidden-layer neural networks trained demonstrates proper regression (Figure 2.16, as an example). Hence, a committee of (only) five networks is employed. The committee overall regression is portrayed in Figure 2.17. The networks' initial two-hidden-layer architecture for training contained 12–6 neurons, as the subsystem is more complex.

2.4.5.2 Knock Detection Subsystem

Knock is a detrimental event, resulting from auto-ignition in a pocket of air-fuel mixture outside of the main flame front propagation region. This is mostly because of too early spark advances and partly for cylinder higher temperature. According to [13], knocking occurs when the induction time integral exceeds 1,[xiii] which is the basis for GT-Power software. The function of knock sensor is to

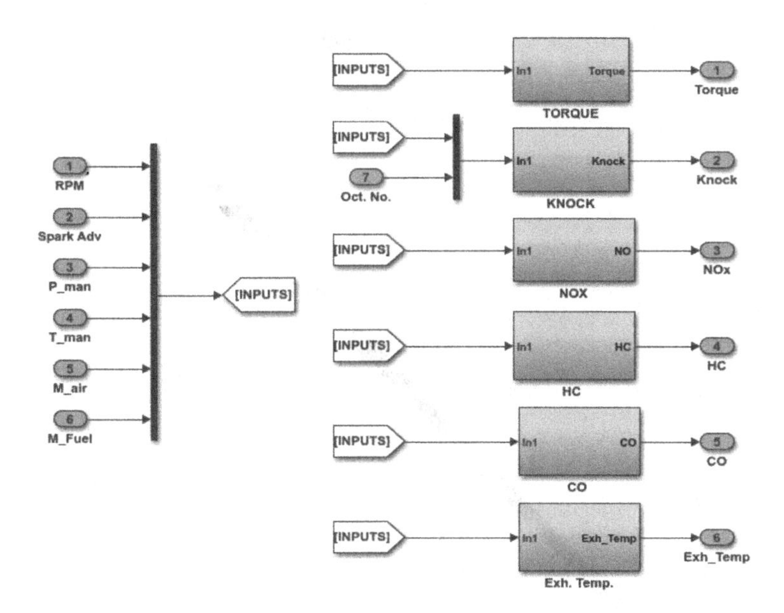

FIGURE 2.15 Combustion subsystem structure.

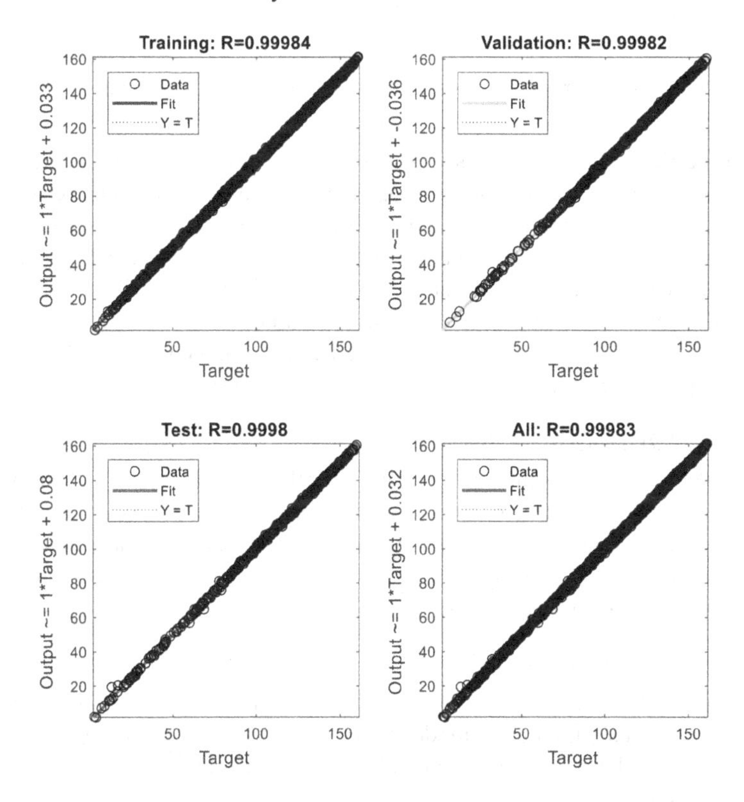

FIGURE 2.16 The regression of a network employed for torque generation subsystem.

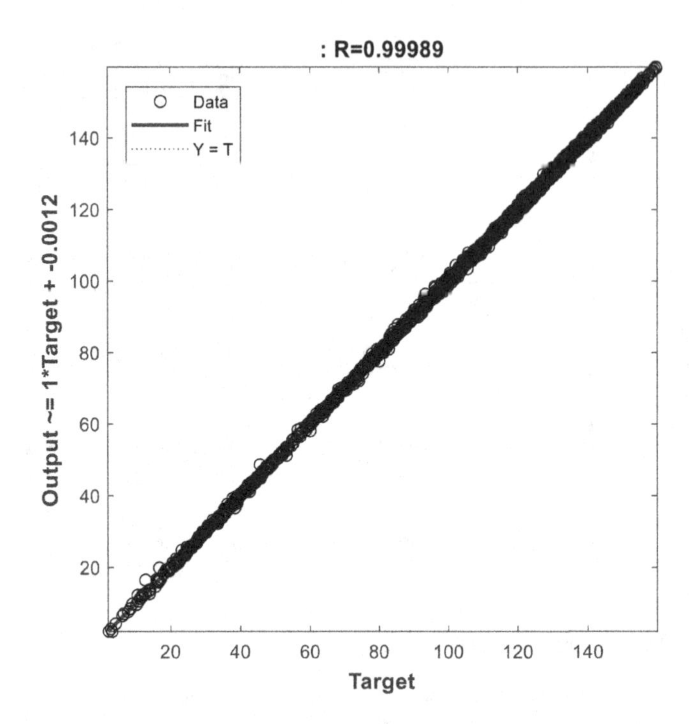

FIGURE 2.17 The committee overall regression of the torque generation subsystem.

inform the ECU if knock occurs in a cylinder. The ignition control module uses this signal to retard the ignition timing. Inasmuch as the final knock signal is either one or zero (meaning the existence or nonexistence of knock, respectively), pattern recognition networks are utilized. The overall confusion chart for a committee of five networks is illustrated in Figure 2.18. As can be seen, only two cases (0.1% of the relevant cases) with outputs valued 1 are wrongly predicted 0, and only 1 case with output 0 is wrongly predicted 1. On the whole, 99.9% of total cases are predicted correctly.

2.4.5.3 NOx Emission Subsystem

As stated before, emissions are modeled in both ppm and g/kW-h. The overall regression for a committee of five networks, for NOx prediction in ppm, is displayed in Figure 2.19. As can be seen, the regression accuracy is notable. The networks' initial two-hidden-layer architecture for training contained 12–8 neurons, as the subsystem is more complex.

2.4.5.4 CO Emission Subsystem

Compared to the other emissions, CO is a less complicated task for identification, and the networks' initial two-hidden-layer architecture for training contained 12–6 neurons. The regression for one of the five networks employed in the committee is displayed in Figure 2.20.

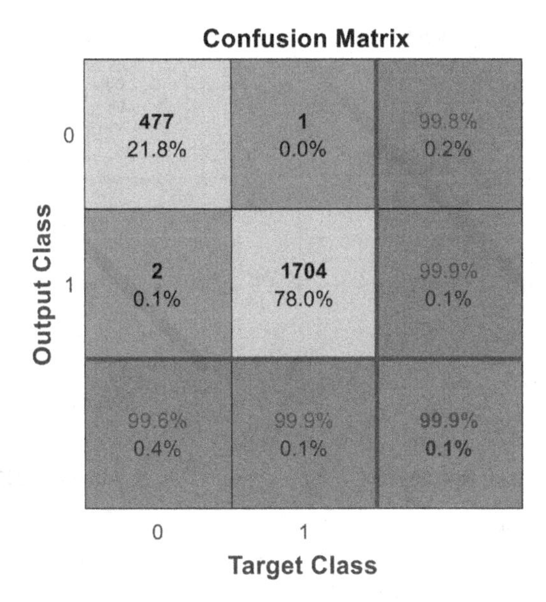

FIGURE 2.18 The committee overall confusion of the knock detection subsystem. Note that overall confusion includes training, validation, and test data confusions all together.

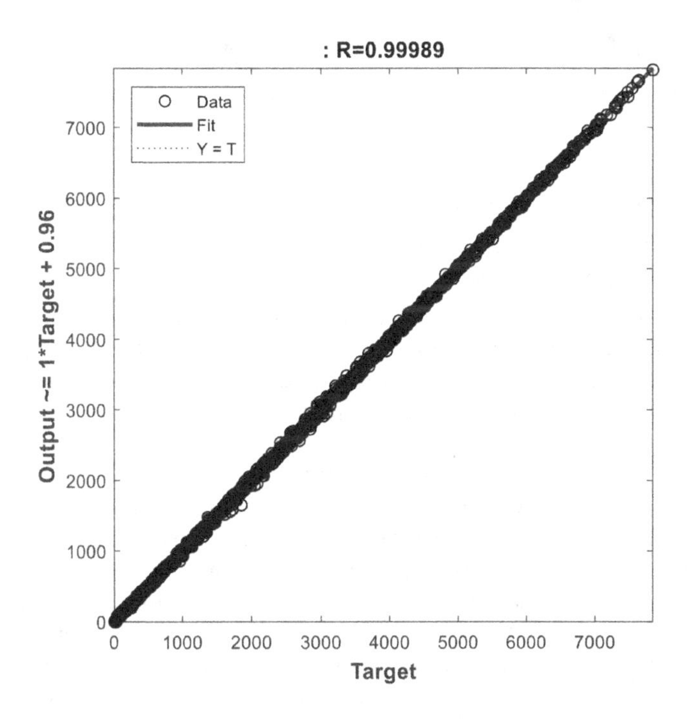

FIGURE 2.19 The committee overall regression of the NOx subsystem in ppm.

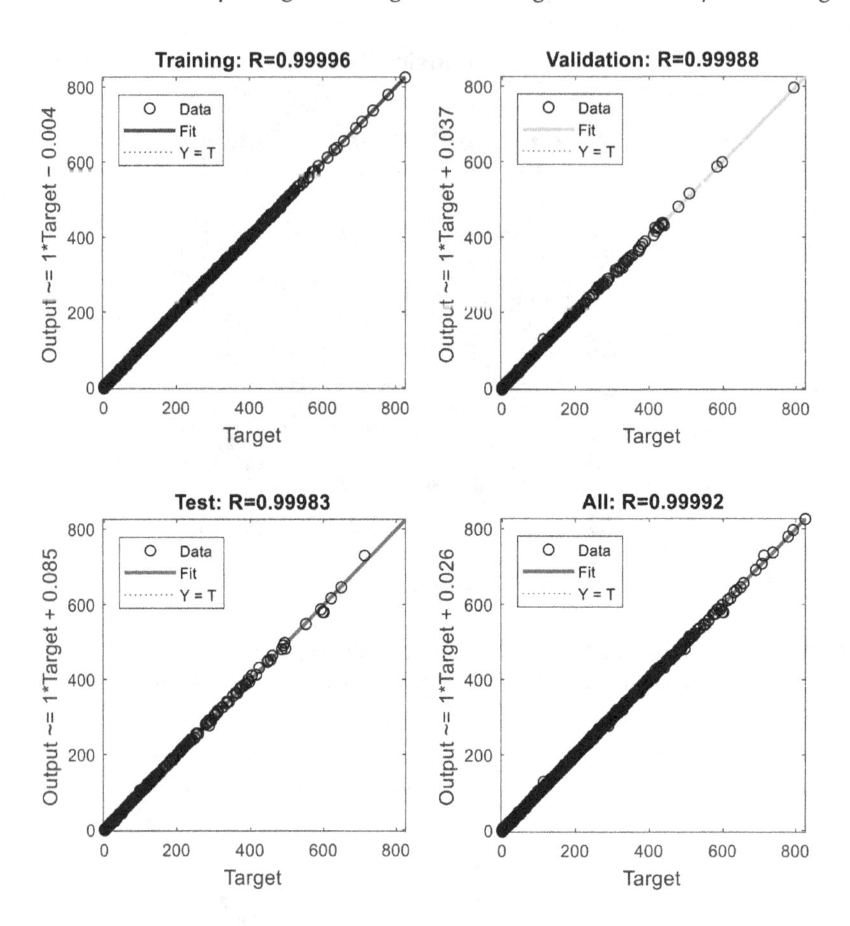

FIGURE 2.20 The regression of a network employed for CO committee in g/kW-h.

2.4.5.5 HC Emission Subsystem

For this most complicated emission, a committee of 10 networks is utilized, and MSE is reduced from 169.34 to 108.71. The committee overall regression is shown in Figure 2.21. The networks' initial two-hidden-layer architecture for training contained 15–10 neurons, as the subsystem is more complex.

2.4.5.6 Exhaust Gas Temperature Subsystem

Exhaust gas temperature is an important constraint in the control system design. For the appropriate operation of the three-way catalytic converter (TWC), it needs to be kept within a specific band: above light-off temperature (for TWC to start working properly) and below maximum allowable temperature (so as not to damage the TWC material). Figure 2.22 depicts the overall regression of the subsystem. Values are in °K. Since the GT-Power reference model might be so accurate in extreme cases,[xiv] too low and too high temperatures are not shown. Yet, the methodology would be the same.

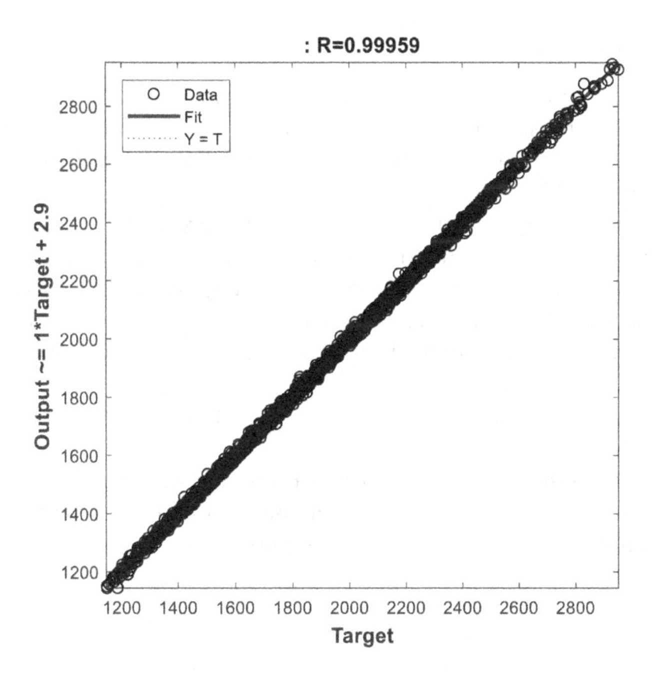

FIGURE 2.21 The committee overall regression for HC subsystem in ppm.

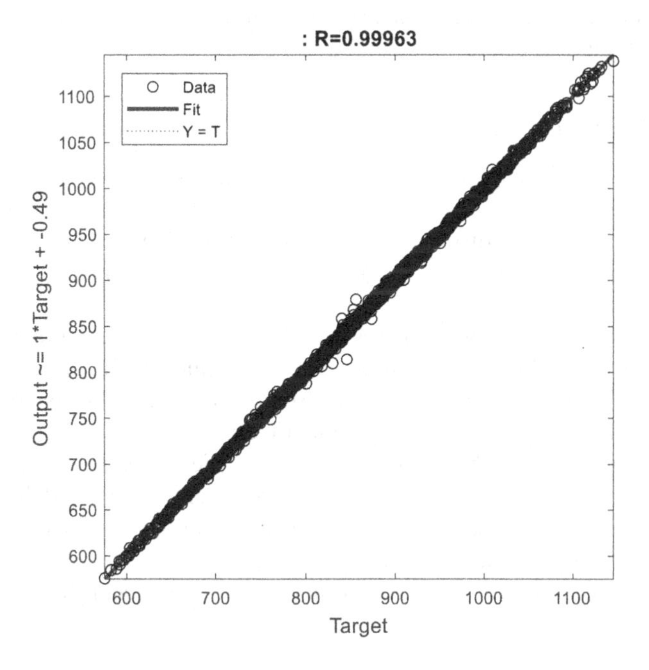

FIGURE 2.22 The overall regression of exhaust gas temperature subsystem. The values are in °K. Too low and too high temperatures are not shown.

2.4.6 Exhaust Manifold

For a more accurate engine model, exhaust manifold could also be taken into account. This subsystem is dynamic and could be modeled just like the intake manifold (Sections 2.4.1 2.4.3). However, here the focus is on thermal behavior of exhaust manifold.

Therefore, the coupled equations (2.8) and (2.9) are again employed to determine mass and, particularly, thermal accumulation in the exhaust manifold. The only difference is that an additional term for enthalpy loss needs to be considered, which is the sum of the heat transfer from the exhaust gases to the exhaust manifold walls and the enthalpy of the flow out of manifold. Namely [6]:

$$\dot{H}_{\text{out}}(t) = \dot{m}_{\text{out}}(t) \cdot c_p \cdot \vartheta_{\text{out}}(t) + \dot{Q}_{\text{in}}(t), \tag{2.15}$$

where $\dot{Q}_{\text{in}}$ denotes the heat transfer from the exhaust gases to the exhaust manifold walls and the enthalpy of the flow out of manifold. A part of this heat is absorbed by the manifold the rest is transferred to the surroundings:

$$m_w(t) \cdot c_w \cdot \frac{d}{dt} \vartheta_w(t) = \dot{Q}_{\text{in}}(t) + \dot{Q}_{\text{out}}(t), \tag{2.16}$$

where index "w" denotes the manifold wall. The heat transfer terms to and from manifold, i.e. $\dot{Q}_{\text{in}}(t)$ and $\dot{Q}_{\text{out}}$, consist of convection and radiation terms, which per se, could be modeled based on heat transfer references, like [14]. A simplified modeling discussion can be found in [6].

2.4.7 Turbocharger

As stated earlier, the engine employed for case study is naturally aspirated (NA). The torque produced in an NA engine is restricted by its volumetric efficiency, or in other words, the displacement volume. Turbochargers are employed to increase the amount of air induced into cylinders, and thus, to raise the produced power.

Turbocharger consists of a compressor (upstream of intake manifold) and a turbine (downstream of exhaust manifold) with a common shaft. Utilizing exhaust gas energy, the turbine rotates the compressor, boosting air induction into engine. This practically raises the effective volumetric efficiency.

Accordingly, three subsystems are added to the system block diagram in Figure 2.1: compressor subsystem – before the receiver (intake manifold dynamic) subsystem; turbine subsystem – after the exhaust manifold subsystem; and turbocharger inertia – after the turbine subsystem.

The turbocharger inertia is modeled just like the rotational dynamic subsystem. That is

$$\Theta_{tc} \frac{d}{dt} \omega_{tc}(t) = T_t(t) - T_c(t) - T_f(t), \tag{2.17}$$

where Θ_{tc} denotes the rotational inertia, ω_{tc} stands for the shaft speed, T_t and T_c represent the turbine and compressor torques, respectively, and T_f is friction loss torque.

The key assumption in turbocharger modeling is that the dynamic phenomena in the fluid-dynamic processes (in both turbine and compressor) are much faster than the variations in thermodynamic boundary conditions. Accordingly, compared to other subsystems, compressor and turbine could be considered *static* [6].

Just as before, static neural networks could be utilized for the compressor and turbine modeling.

Figure 2.23 portrays turbocharger subsystems with their layout in the system's block diagram, with respect to intake and exhaust receivers. Note that a turbocharger is typically triggered at full-load.

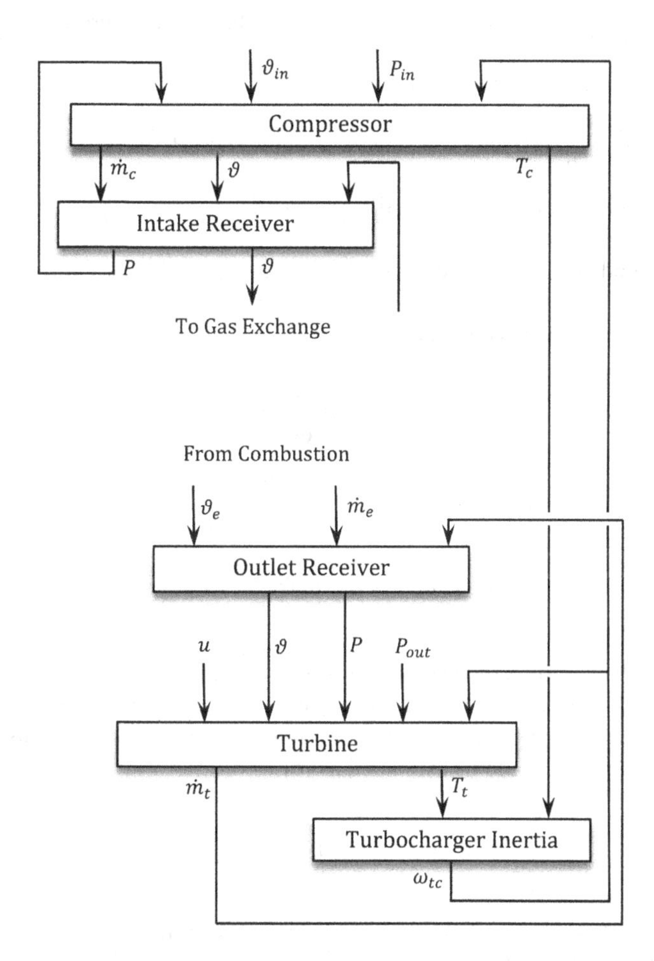

FIGURE 2.23 Turbocharger subsystems with their layout in the system's block diagram, with respect to intake and exhaust receivers [6].

2.4.7.1 Compressor

The inputs to this subsystem are the inlet and outlet pressures (or inlet–outlet pressure ratio, Π_c), inlet temperature ϑ_{in}, and rotational (shaft) speed ω_{tc}. The outputs include mass flow through the compressor $\dot{m}_c$ and the torque T_c required to rotate the compressor (for the given states).

Ordinarily, performance curves, as compressor maps, are employed to estimate the mass flow rate $\dot{m}_c$ and also, the respective isentropic efficiency η_c. These maps are typically represented with respect to corrected variables, which are given as follows:

- $\Pi_c = \dfrac{P_{\text{out}}}{P_{\text{in}}}$; pressure ratio
- $\tilde{\omega}_c = \sqrt{\dfrac{\vartheta_{\text{in},0}}{\vartheta_{\text{in}}}} \cdot \omega_{tc}$; corrected speed ($\vartheta_{\text{in},0}$ is the inlet temperature at which the map was originally measured)
- $\dot{\mu}_c = \dfrac{P_{\text{in}}}{P_{\text{in},0}} \sqrt{\dfrac{\vartheta_{\text{in},0}}{\vartheta_{\text{in}}}} \cdot \dot{m}_c$; corrected mass flow rate ($P_{\text{in},0}$ is the inlet pressure at which the map was originally measured)

As mentioned, $\dot{m}_c$ and η_c are derived with respect to Π_c and $\tilde{\omega}_c$, in the compressor map.

Compressor torque is computed as

$$T_c = \frac{P_c}{\omega_{tc}}, \tag{2.18}$$

where the compressor power, P_c, is estimated as

$$P_c = \dot{m}_c \cdot c_p \cdot \vartheta_{\text{in}} \cdot \left[\Pi_c^{(k-1)/k} - 1\right] \cdot \frac{1}{\eta_c}, \tag{2.19}$$

where η_c is isentropic efficiency of the respective operating point. Note that $\dot{m}_c$ and η_c are already derived from the compressor map, explained above.

Similarly, assuming isentropic compression and adiabatic conditions, the outlet temperature ϑ_{out} could be estimated as [6]

$$\vartheta_{\text{out}} = \vartheta_{\text{in}} + \left[\Pi_c^{(k-1)/k} - 1\right] \cdot \frac{\vartheta_{\text{in}}}{\eta_c}. \tag{2.20}$$

In our methodology given below, these all (particularly the map) could simply be replaced by neural network structures, for more accurate estimations.

2.4.7.1.1 Limitations

Every compressor is restricted by at least four limiting conditions, resulting in a limited operating range. They are as follows:

- Mechanical limit, the maximum rotational speed, so as to prevent large centrifugal forces
- Choke limit, where the flow in the narrowest part of compressor reaches sonic conditions
- Blocking limit, the behavior of the compressor at very low speeds where it acts like a blocking orifice
- Surge limit, where fluid-dynamic instability occurs, destroying the regular flow patterns inside the compressor [6]

2.4.7.1.2 Extrapolation

There is a small empty region in compressor map (pertaining to low load and speeds below 10,000 rpm), where speed measurement is difficult (and thus, is avoided). In this region, mass flows and the pressure ratios are small, leading to large relative errors. Accordingly, extrapolation fails to yield acceptable results.

To resolve this problem, dimensional analysis is utilized. To do so, non-dimensional numbers including the flow coefficient (Φ, aka, normalized compressor flow rate), circumferential Mach number (Ma), and isentropic work coefficient (Ψ, aka, head parameter), are defined as follows:

$$\Psi = \frac{c_p \cdot \vartheta_{\text{in}} \cdot \left[\Pi_c^{(k-1)/k} - 1\right]}{\frac{1}{2} U_c^2},$$

$$\Phi = \frac{m_{c,\text{cor}}}{\frac{\pi}{4}\,\rho_{\text{in}}\, d_c^2\, U_c},$$

$$\text{Ma} = \frac{U_c}{\sqrt{k\, R\, \vartheta_{\text{in}}}} \tag{2.21}$$

where U_c and $m_{c,\text{cor}}$ are corrected speed and mass flow rate, and d_c is the diameter of compressor blade.

Using these non-dimensional factors, extrapolating functions could be defined, and the turbocharger maps can cover the whole operational range [6, 11].

2.4.7.2 Turbine

In control perspective, flow dynamic behavior in turbine could be assumed like an orifice (with a variable area in the case of variable geometry turbines [VGTs]), quite the same methodologies as compressors hold for turbines. The inputs are inlet and outlet pressures, the shaft speed, and inlet temperature, along with an extra input as control command. This command could be waste-gate position or turbine blade angles (for VGTs). The outputs include mass flow rate through turbine and turbine torque.

Just as compressors, performance curves as turbine maps are provided with respect to corrected variables. The corrected mass flow rate μ_t is derived based on pressure ratio Π_t and corrected speed $\tilde{\omega}_t$ from the turbine map. From another map, the turbine efficiency is derived based on pressure ratio, and a new parameter, that is, turbine blade speed ratio $\tilde{c}_{us}$:

$$\tilde{c}_{us} = \frac{r_t \cdot \omega_t}{c_{us}}, \tag{2.22}$$

where

$$c_{us} = \sqrt{2\, c_p \cdot \vartheta_{in} \cdot \left[1 - \Pi_t^{(1-k)/k}\right]}. \tag{2.23}$$

Turbine torque is computed as

$$T_t = \frac{P_t}{\omega_{tc}}, \tag{2.24}$$

in which, the turbine power, P_c, is estimated as [6]

$$P_t = \dot{m}_t \cdot c_p \cdot \vartheta_{in} \cdot \left[1 - \Pi_t^{(1-k)/k}\right] \cdot \eta_t. \tag{2.25}$$

Again, in the following of our methodology, these all (particularly the maps) could simply be replaced by neural network structures, for more accurate estimations.

2.4.8 Engine Thermal Model

As explained before, engine temperature varies very slowly. In other words, the time constant for engine thermal behavior is large. Recall that subsystems with very large time response are regarded constant (Section 1.3). Consequently, this subsystem could be assumed constant. Using a simplified lumped-parameter modeling approach, engine thermal balance could be expressed as

$$m_e \cdot c_e \cdot \frac{d}{dt} \vartheta_e(t) = \dot{H}_w(t) + \alpha \cdot A \cdot \left[\vartheta_e(t) - \vartheta_a\right], \tag{2.26}$$

where m_e represents the engine mass, c_e denotes the engine-average-specific heat capacity, $\dot{H}_w(t)$ is the heat transfer to the engine block, α represents the convective heat transfer coefficient, A is the active heat-exchange area, and ϑ_a denotes the ambient air temperature.

Now, assuming engine efficiency $\eta_e(t)$, and $\dot{H}_{eg}(t)$ as the enthalpy exiting the system with the exhaust gases, $\dot{H}_w(t)$ could be expressed as

$$\dot{H}_w(t) = \left(1 - \eta_e(t)\right) \cdot H_l \cdot \dot{m}_\varphi + \dot{H}_{eg}(t), \tag{2.27}$$

where $\dot{m}_\varphi$ is the injected fuel mass flow rate, and H_l denotes the fuel's lower heating value. Therefore, the engine's thermal time constant (τ_e) could be derived as

$$\tau_e = \frac{m_e \cdot c_e}{\alpha \cdot A}, \tag{2.28}$$

which would be about several thousand seconds [6].

A more detailed engine thermal modeling could be found in [15] or [6].

2.4.9 Catalytic Converter Subsystem

TWC is one of the important subsystems with crucial impact on calibration. In fact, the final engine model would be employed for the design of software-in-the-loop (SiL) or MiL control maps, which, per se, demand after-catalyst emissions for a more sensible design.

For this subsystem, already prepared MATLAB catalyst models can be employed. In addition, TWC control-oriented models could also be utilized. A thorough discussion in this regard can be found in [6].

Here, as an alternative, a TWC modeling approach via neural networks is proposed.

The outputs of this subsystem involve extensive intervals of values from extremely high to extremely low (even zero), making this subsystem probably the most challenging one for identification. It shall be noted that it is almost impossible for one individual neural network to learn both very large and very small values at the same time.[xv] As a result, global-local model structures such as ANFIS, LOLIMOT, LOPOMOT (i.e. local linear polynomial model tree), etc., could be employed.

Another recourse which brings more flexibility is to use (our method as) improved partitioning[xvi] (Section 2.2). In this way, each partition could comprise different committees, making it more flexible for our application.

Like before, each emission output is identified by one separate subsystem. The inputs include combustion emissions in gram/second (derived from emission models with outputs in g/kW-h, and multiplying them by the produced power), and also the incoming flow temperature. Based on physical insight (which is computationally proven in our investigations), "Air-fuel ratio" input is also added as an assisting knowledge signal, to facilitate, and thus, to improve the identification.

First, a partitioning is performed based on the exhaust flow temperature input. In other words, since catalytic converter shows totally different behaviors for exhaust flow input below light-off temperature (compared to its normal operation), two separate catalyst network models are designed for these two regimes, one for high temperature exhaust input and another for low temperature exhaust input. Just as mentioned before, the area around the border between the two regimes is included in the pattern tables of both models to improve the overall prediction quality. Figure 2.24 illustrates the respective partitioning.

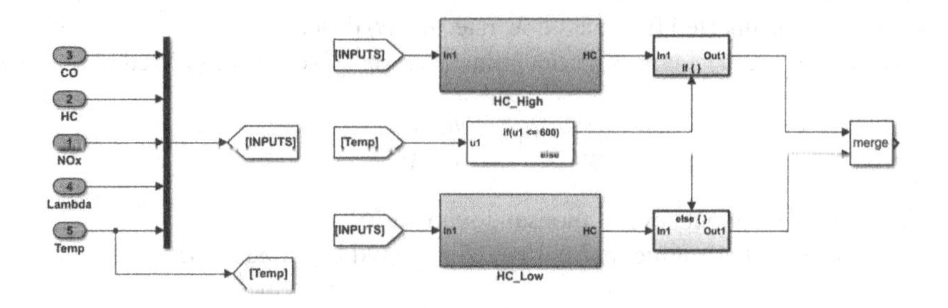

FIGURE 2.24 Partitioning based on the exhaust flow temperature input.

Although simpler, the tasks are not still easy for identification since the pattern tables still involve both very high and very low values.

Interestingly, for some intervals of air-fuel ratio, catalyst emission outputs become much lower than the emission standard limits. In these intervals, using an engineering insight, the respective output can be assumed zero, making it an ideal interval to be assigned as one partition. The remaining air/fuel input interval would only contain larger output values, yielding a simpler task for identification. In accordance with our approach (improved partitioning), for better accuracy, the training data for the network partitions can have overlap with neighboring partitions.

On the whole, this simple, yet interesting, partitioning strategy (Figure 2.25) yields much more accurate results.

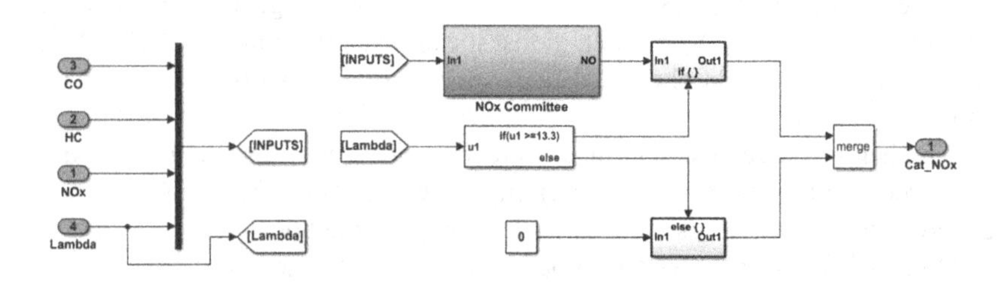

FIGURE 2.25 The architecture of partitioning for catalyst NOx subsystem. The extra exhaust temperature input could be added, particularly for the low-temperature partition.

The behavior of catalytic converter is somewhat dynamic. However, since the engine in hand does not have after-catalyst lambda sensor, static modeling quite suffices for the controller design.[xvii] Besides, as will be seen in the results, static modeling of catalyst yields decent accuracies. For a more accurate modeling, dynamic neural networks could be utilized.

In what follows, catalyst emission subsystems are designed. Just as before, ensemble averaging is employed. For these subsystems, due to gradient considerations, “Conjugate Gradient with Beale-Powell Restarts” training algorithm is applied. The networks’ initial two hidden-layer architecture for training contains 15–10 neurons.

2.4.9.1 Catalyst NOx Emission Subsystem

The architecture of partitioning for Catalyst NOx subsystem is depicted in Figure 2.25. A committee of 10 networks is incorporated in the upper partition. Figure 2.26 displays the overall regression for the best individual network trained without partitioning, while Figure 2.27 indicates the overall regression for the proposed structure. As can be seen, the identification quality has remarkably improved. The MSE is reduced from 4.504e−12 (for the best individual network) to 4.017e−13 (for the proposed structure), namely, more than 10 times reduction.

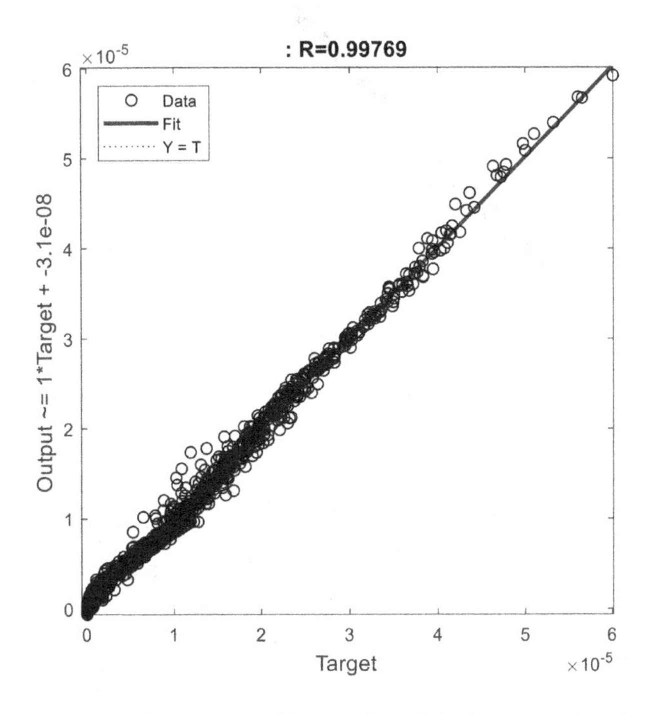

FIGURE 2.26 The unfavorable overall regression of the best network trained for NOx without partitioning in kg/s.

Figures 2.28 and 2.29 re-display the aforesaid figures, in a smaller scale close to zero, which is the bone of contention. As can be seen, the proposed method is well capable of identifying smaller values.

2.4.9.2 Catalyst CO Emission Subsystem

Since CO is slightly less complicated than the other two emissions, a committee of (only) five networks is employed for the lower partition, while upper partition is set to zero. Figure 2.30 demonstrates the proper overall regression of the subsystem. The MSE is decreased from 2.531e−10 (as for the best individual neural network trained without partitioning) down to 2.475e−11, which shows more than 10 times reduction.

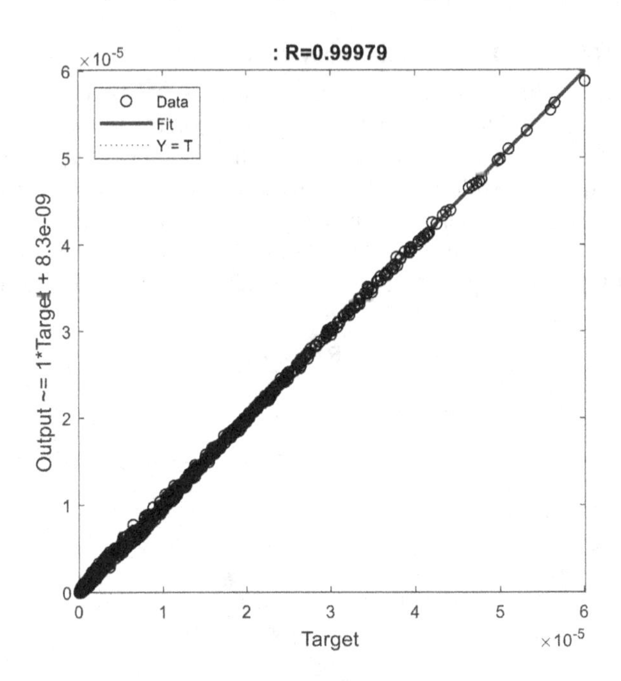

FIGURE 2.27 The overall regression for catalyst NOx subsystem with the proposed structure in kg/s.

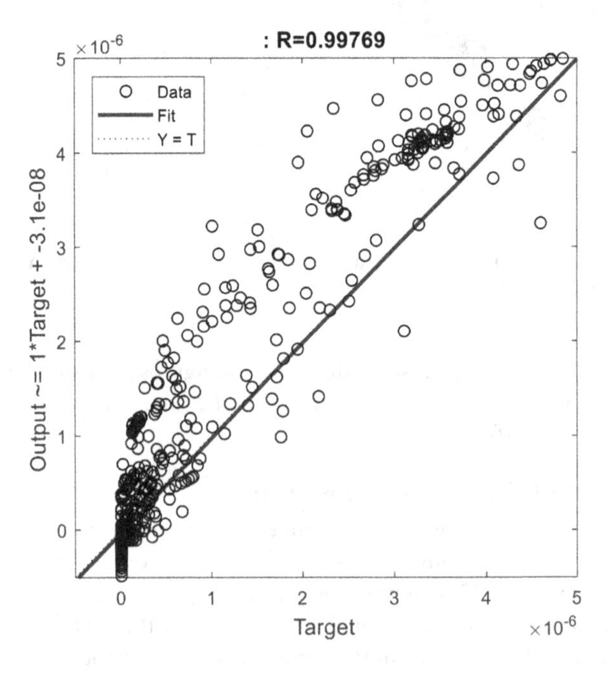

FIGURE 2.28 Re-scaling of Figure 2.26. The adverse regression of the best individual network trained as for smaller values.

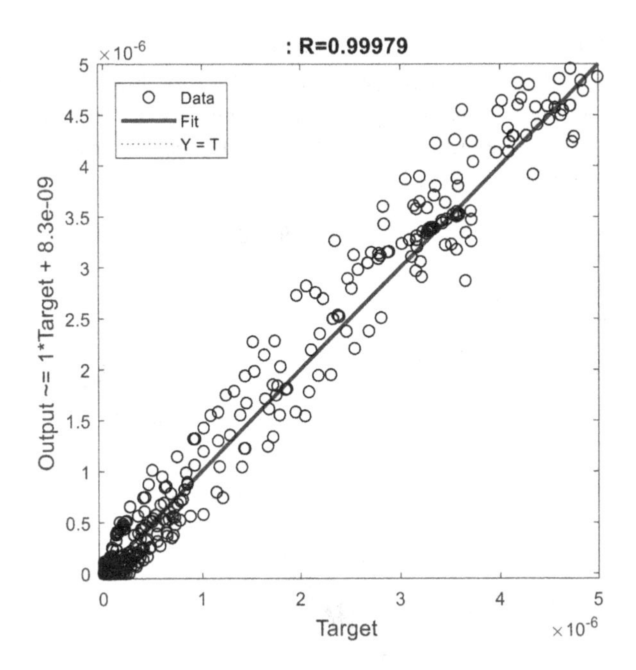

FIGURE 2.29 Re-scaling of Figure 2.27. The favorable overall regression of the proposed structure as for smaller values.

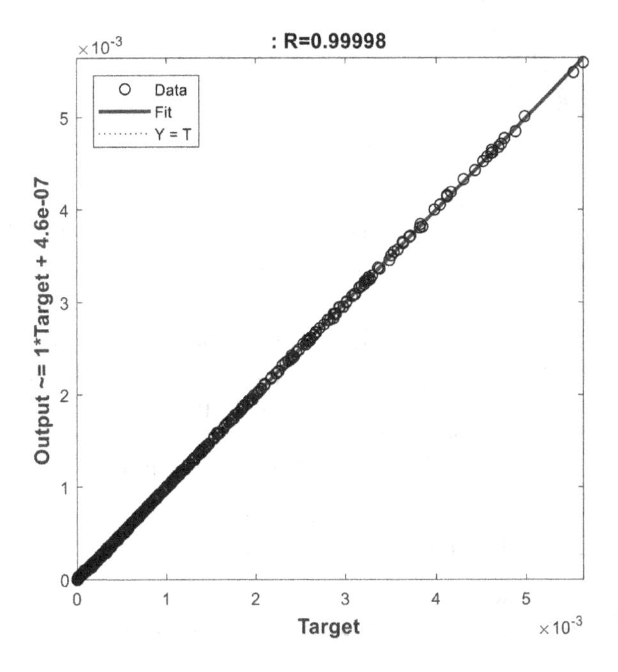

FIGURE 2.30 The overall regression of the proposed structure as for catalyst CO subsystem in kg/s.

2.4.9.3 Catalyst HC Emission Subsystem

A committee of 10 neural networks is employed for the lower partition. MSE is reduced from 8.761e−12 (as for the best individual network trained without partitioning) down to 1.317e−12 for the proposed structure. The overall regression is depicted in Figure 2.31.

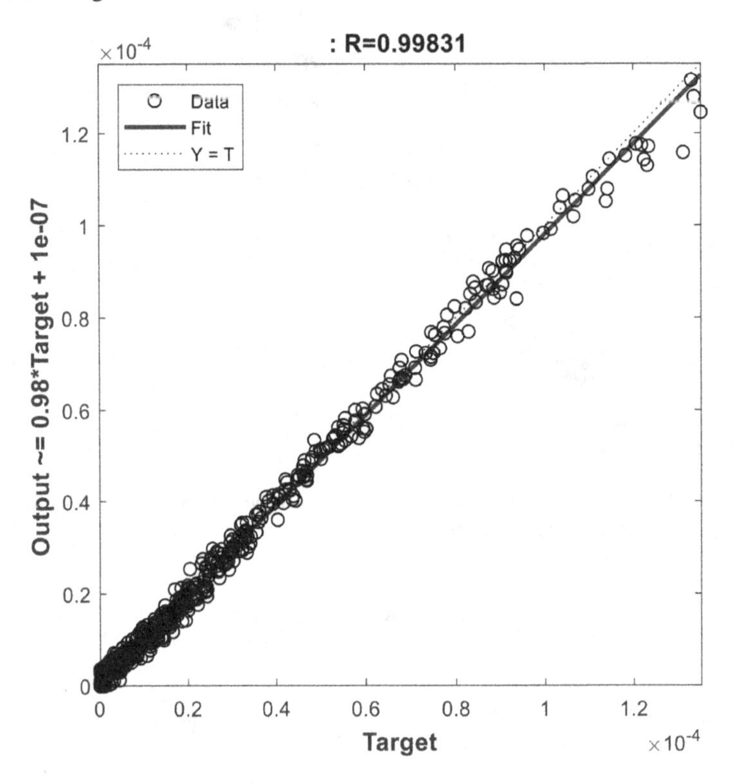

FIGURE 2.31 The overall regression of the proposed structure as for catalyst HC subsystem in kg/s.

It is worth mentioning that HC and NOx emissions (according to their physical behavior) could also be divided into three partitions: the middle one pertaining to air-fuel ratios around stoichiometric value, one partition concerning the upper band and one relating to the lower band. However, this would cause less training data for each partition, meaning possibility of more variance error.

2.5 MODEL VALIDATION AND DISCUSSION

In this section, the integrated model (Figure 2.32), which we call Neuro-MVM, is validated based on software-experimental results from GT-Power software (Chapter 1).[xviii]

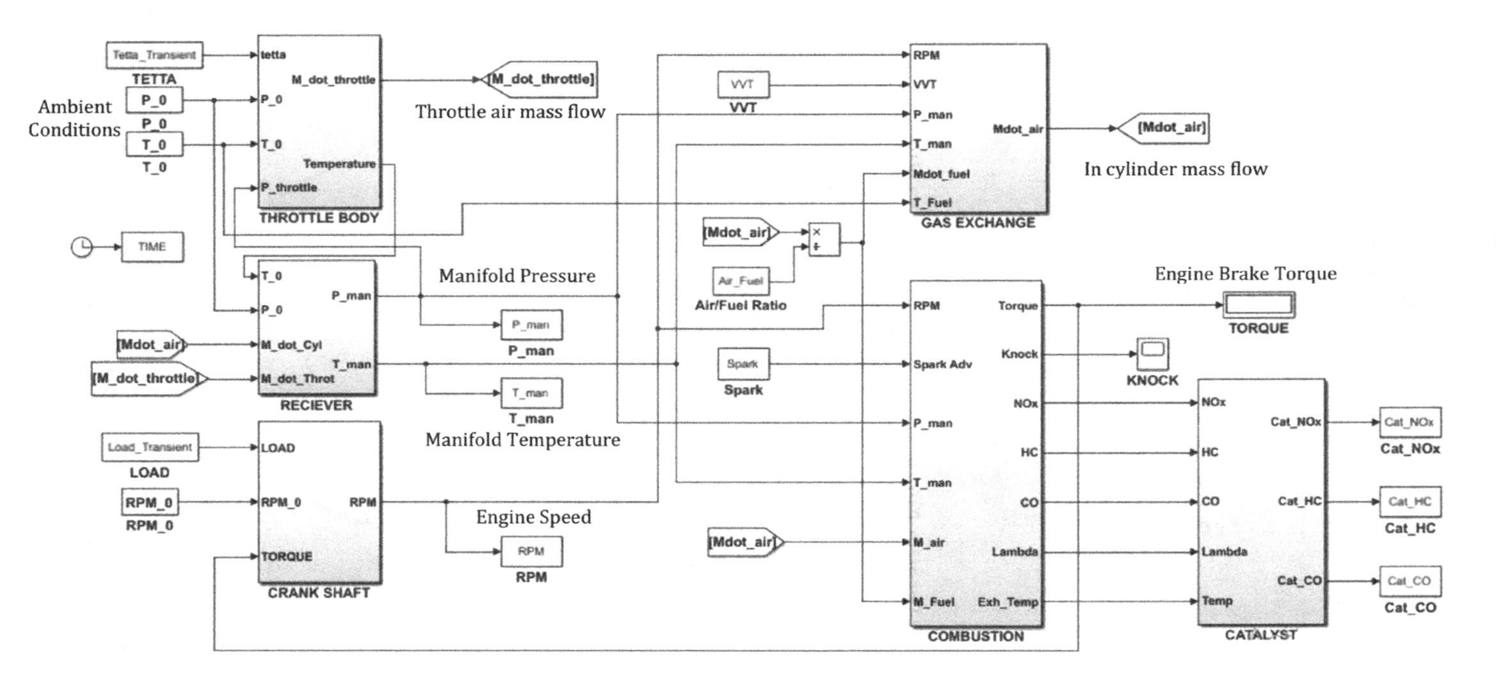

FIGURE 2.32 The schematic of the integrated system model.

The system inputs imposed for ambient circumstances of P_0 = 1 Bar, ϑ_0 = 298 K, and Octane Number = 94 are indicated in Figure 2.33. As can be seen, all the inputs cover widest ranges of operation with sharp variations to challenge the model competence. In fact, the inputs imposed may not seem normal (or even

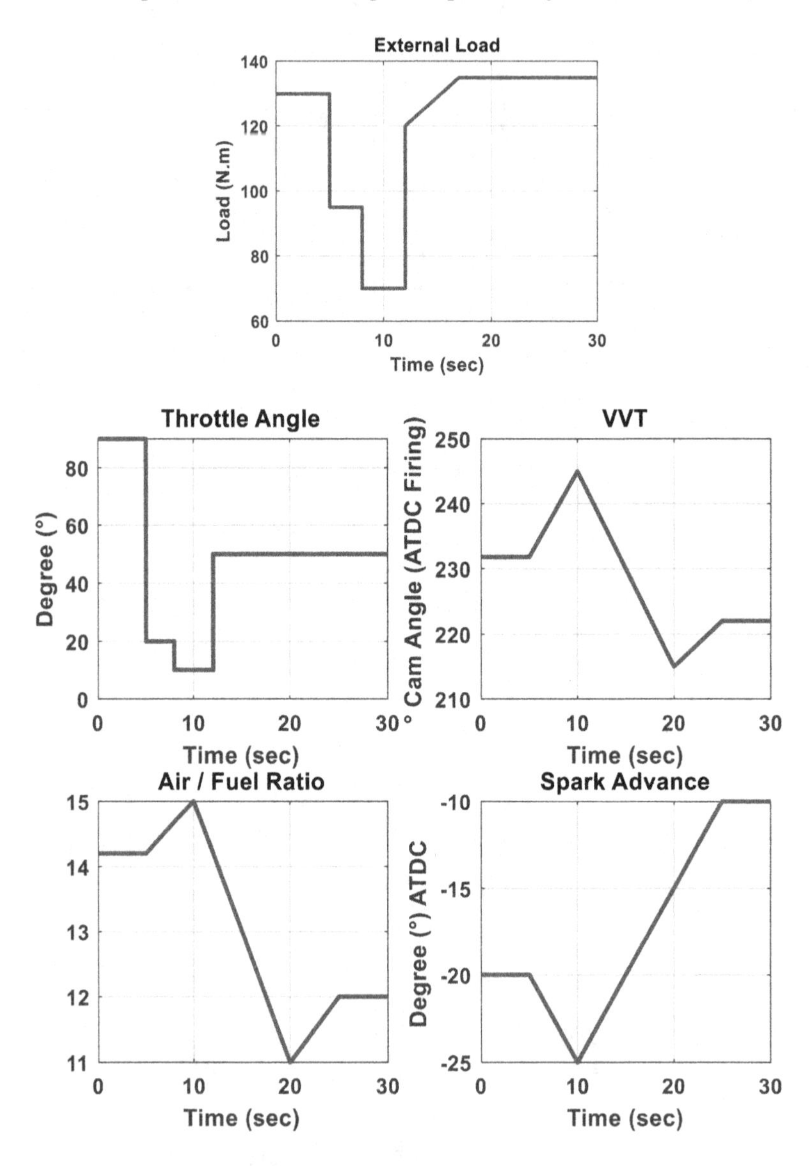

FIGURE 2.33 The system inputs for test. As can be seen, the inputs cover the widest ranges of engine operation. Note that the inputs imposed may not seem normal (or even realistic), yet they are to challenge the model and demonstrate its capabilities, as a control-oriented model is supposed to accurately represent the system's behavior in both favorable and unfavorable conditions.

realistic), yet they are to challenge the model and demonstrate its capabilities, as a control-oriented model is supposed to accurately represent the system's behavior in both favorable and unfavorable conditions. The engine speed, as the most critical output, is compared in Figure 2.34 for both the model and GT test, which signifies a notable resemblance, covering the whole operating region with steady, transient, and transient-to-steady conditions. Since this value is fed back to the system, any error would cause all the other outputs to err. Note also that the inherent noise in test results is filtered for the sake of better representation. Figure 2.35 demonstrates the proper agreement of knock detection in Neuro-MVM and GT test. It is worth reaffirming that knock detection plays a significant role in ignition control.

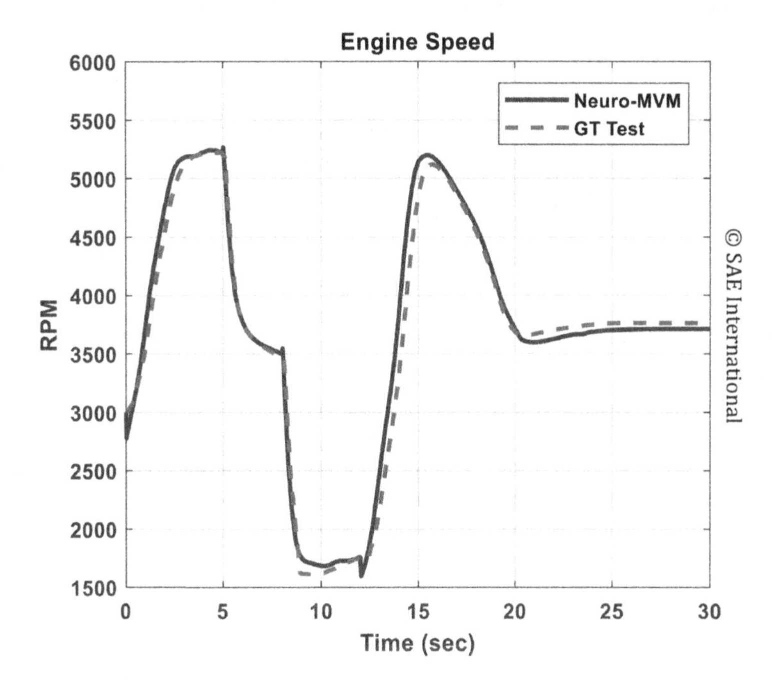

FIGURE 2.34 Comparison of engine speed results for Neuro-MVM and GT test. Standard deviation equals 148.7 rpm.

Figure 2.36 compares manifold pressure in both Neuro-MVM and GT test. The high accuracy of predicting engine emissions is demonstrated in Figures 2.37–2.39.[xix] Brake specific emission results (in g/kW-h) are of the same behavior and are not shown for the sake of brevity.

After-catalyst emissions are displayed in Figures 2.40–2.42, and Figure 2.43 depicts knock index. Unfortunately, only 15 s of GT test simulation is achievable for after-catalyst emissions[xx]; yet, it indicates appropriate accuracy.

Note that the difference in high frequencies (e.g. in an NOx diagram) is not an issue since the GT test results are filtered, and also, it is the area under the curves that would later be deemed in calibration procedure.

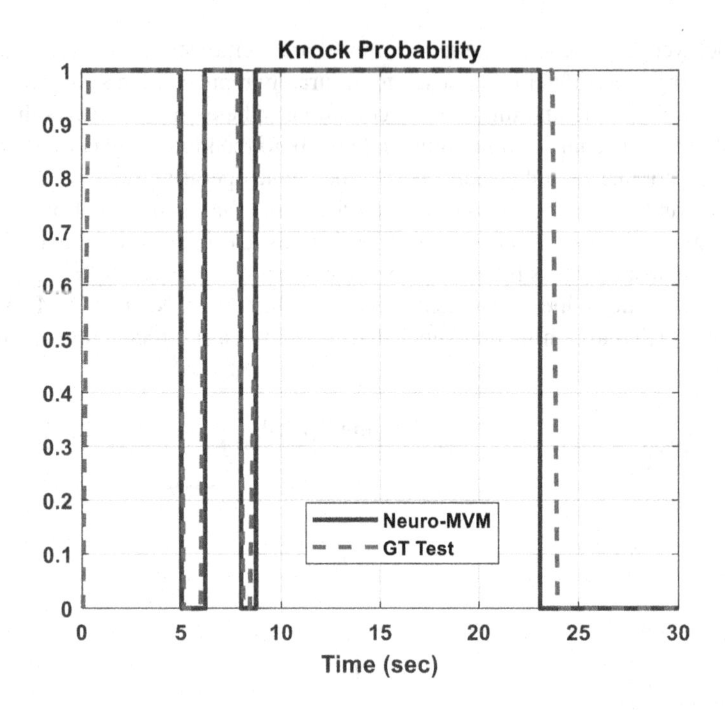

FIGURE 2.35 Comparison of knock detection for Neuro-MVM and GT test.

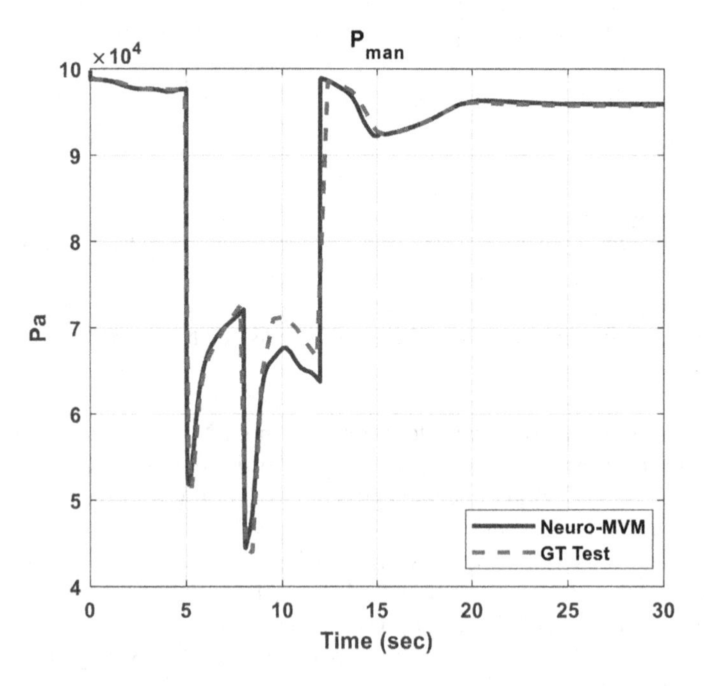

FIGURE 2.36 Comparison of manifold pressure for Neuro-MVM and GT test. Standard deviation equals 2.4066e+03 Pa.

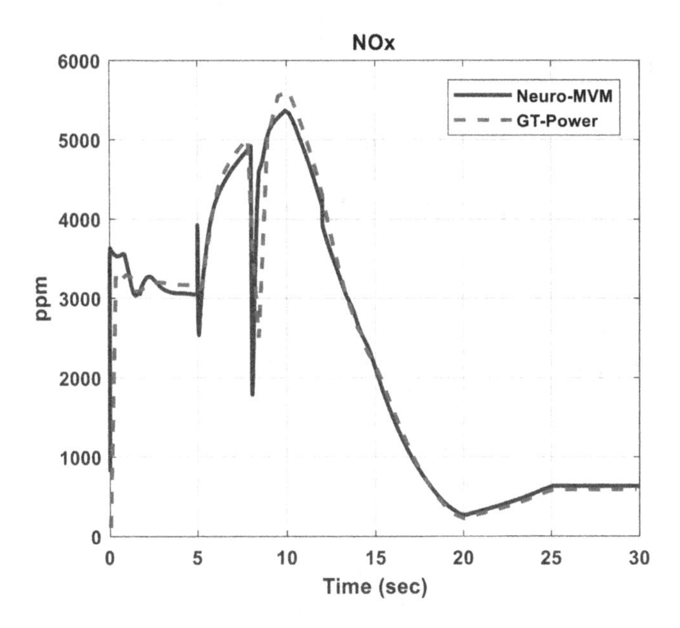

FIGURE 2.37 Comparison of NOx emission results for Neuro-MVM and GT test in ppm. Note that the difference in high frequencies is not an issue since the GT test results are filtered, and also it is the area under the curve that would later be considered in optimization procedure. Standard deviation equals 395.3 ppm.

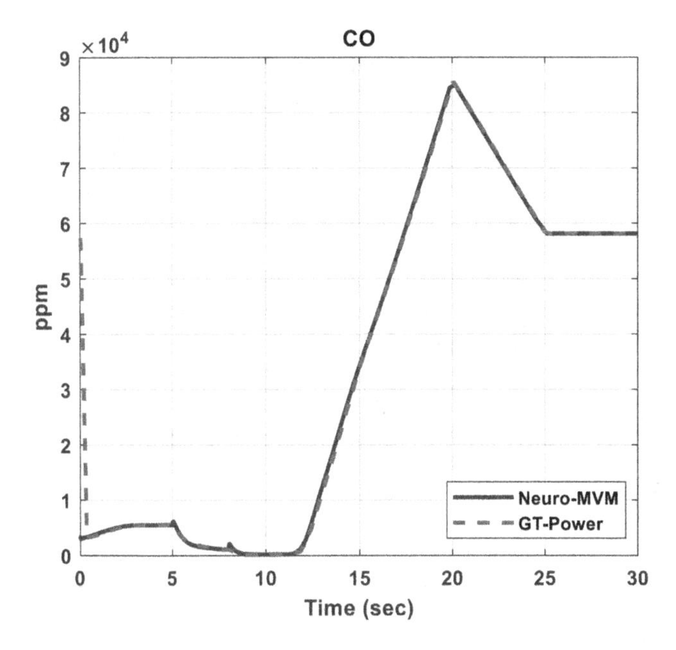

FIGURE 2.38 Comparison of CO emission results for Neuro-MVM and GT test in ppm. Standard deviation equals 4.0401e+03.

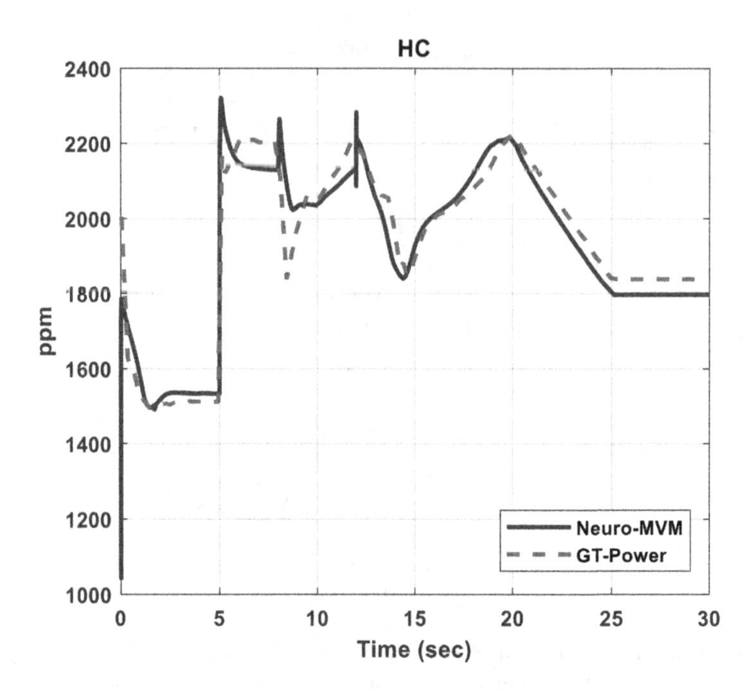

FIGURE 2.39 Comparison of HC emission results for Neuro-MVM and GT test in ppm. Standard deviation equals 57.5852.

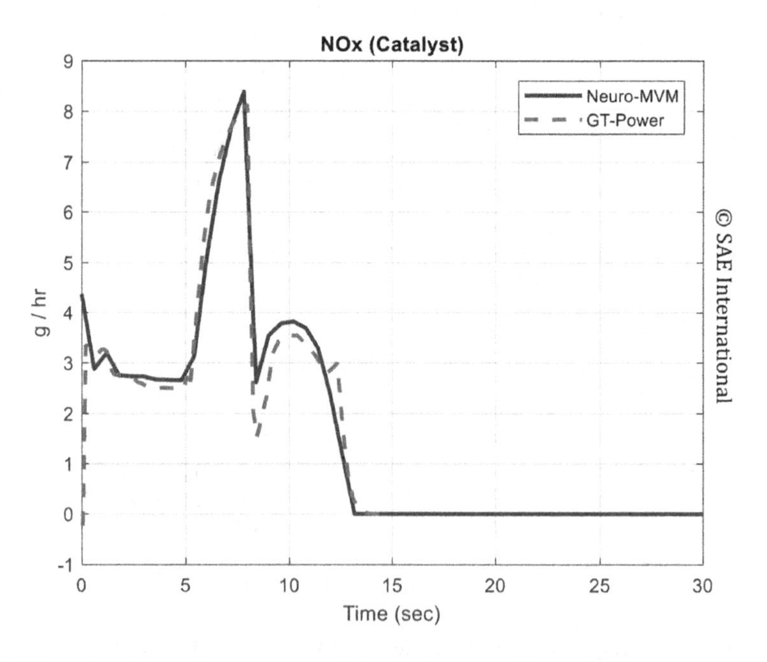

FIGURE 2.40 Comparison of catalyst NOx for Neuro-MVM and GT test.

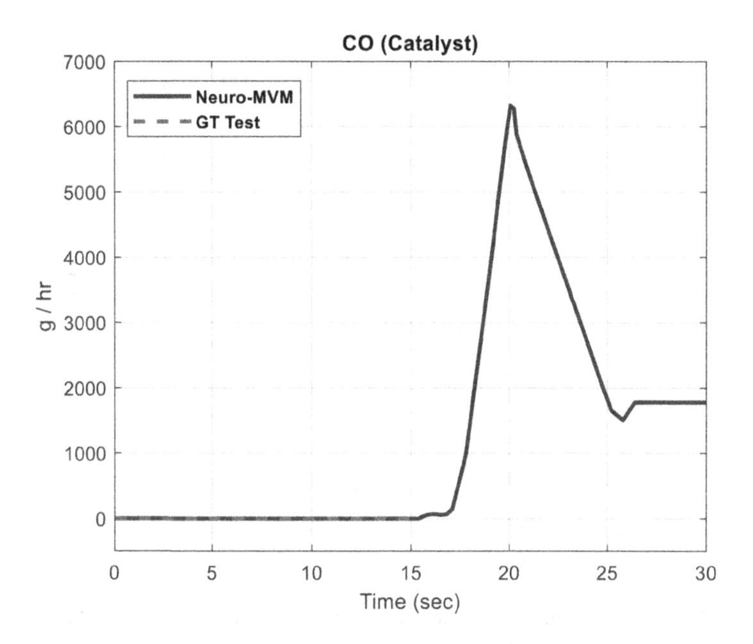

FIGURE 2.41 Comparison of catalyst CO for Neuro-MVM and GT test.

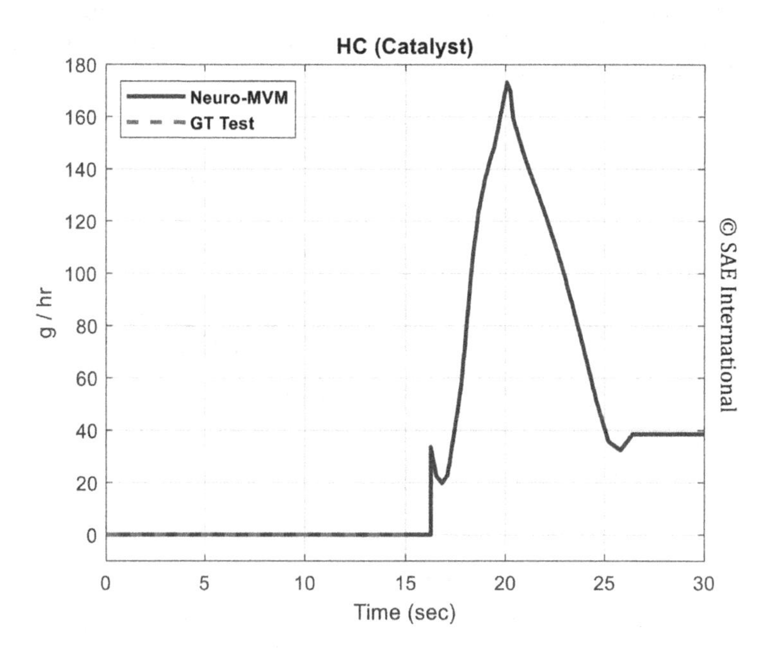

FIGURE 2.42 Comparison of catalyst HC for Neuro-MVM and GT test.

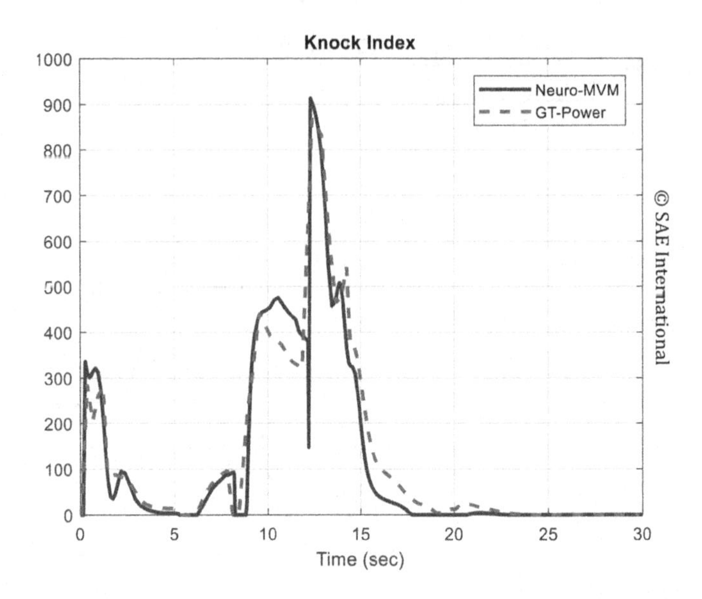

FIGURE 2.43 The comparison of knock index for the model and GT-Power results. Standard deviation equals 50.81.

Now, for ambient circumstances $P_0 = 0.95$ Bar, $\vartheta_0 = 285$ K, and Octane No. = 98, engine speed, knock detection, and NOx emission, as the most crucial outputs, are depicted in Figures 2.44–2.46. Other inputs and outputs are not shown for the sake of brevity.

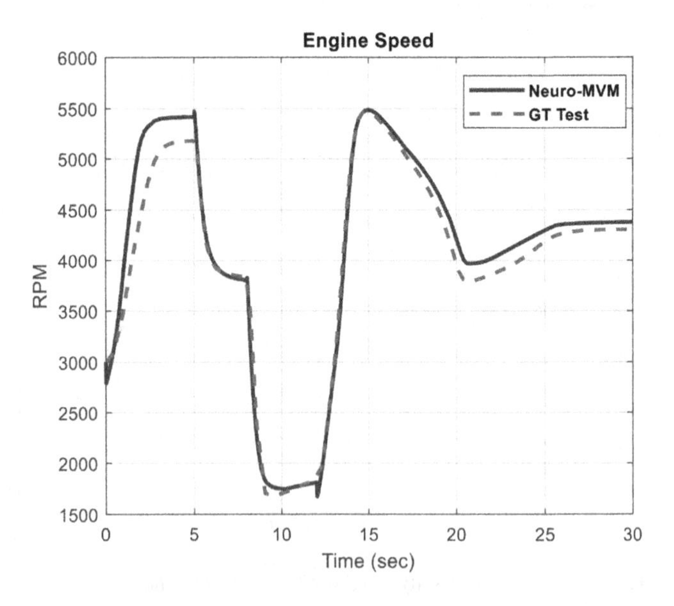

FIGURE 2.44 Comparison of engine speed results for new circumstances.

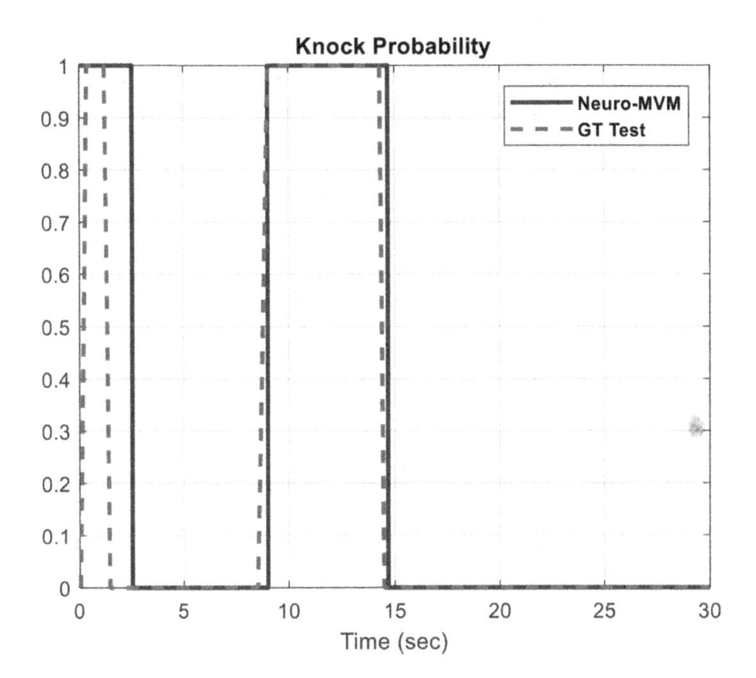

FIGURE 2.45 Comparison of knock results for the new circumstances.

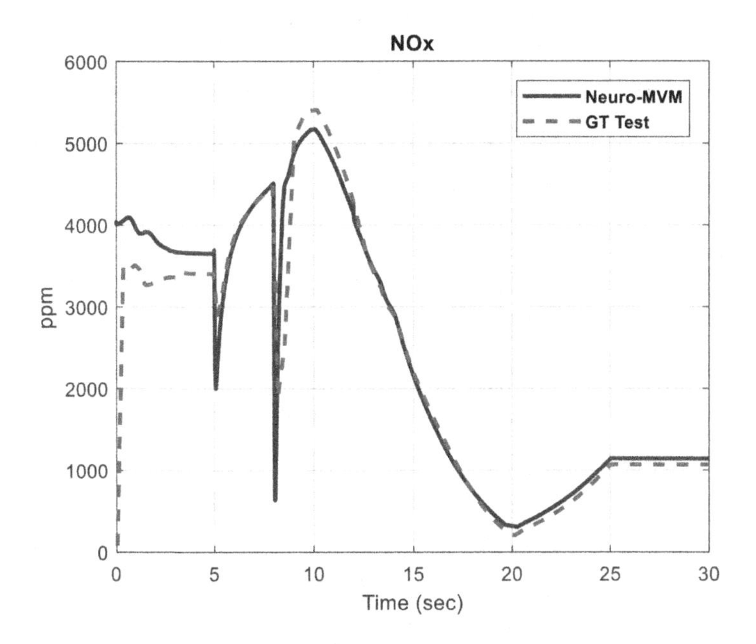

FIGURE 2.46 Comparison of NOx results for the new circumstances.

For ambient circumstances of P_0 = 0.90 Bar, ϑ_0 = 310 K, and Octane No. = 88, engine speed, knock detection, and NOx emission are depicted in Figures 2.47–2.49. Again, other inputs and outputs are not shown for the sake of brevity.

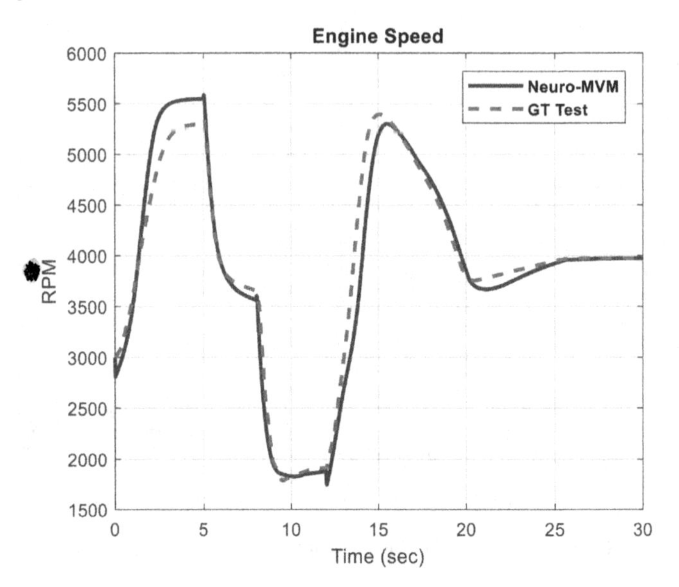

FIGURE 2.47 Comparison of engine speed results for new circumstances.

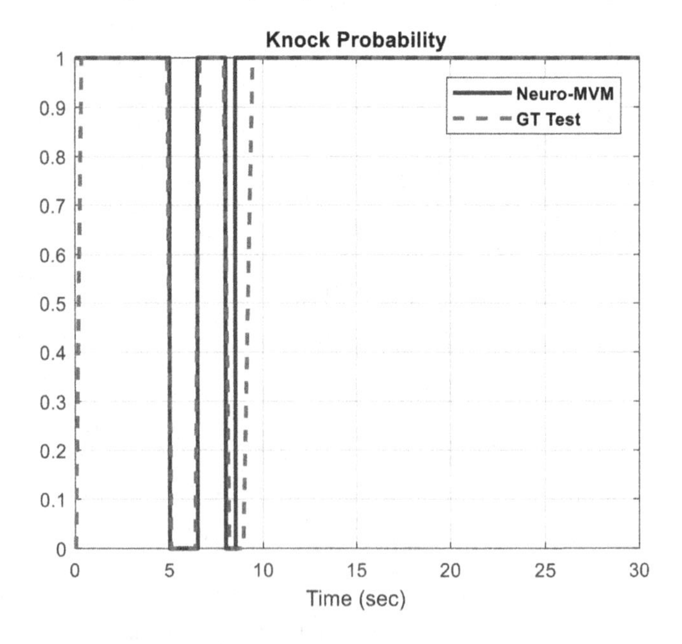

FIGURE 2.48 Comparison of knock results for the new circumstances.

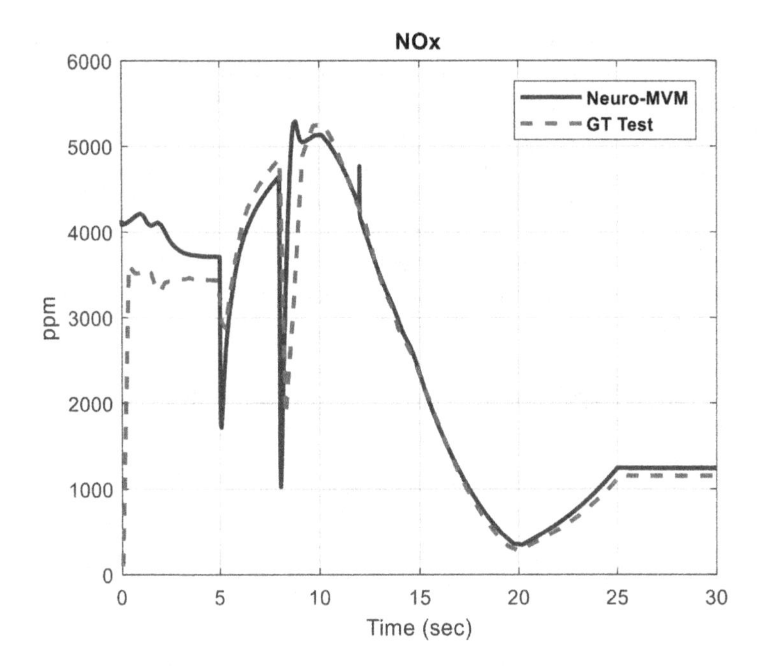

FIGURE 2.49 Comparison of NOx results for the new circumstances.

The combination of gray-box modeling and committee methods is of essential importance in the model success, without which, as demonstrated in our examinations, the model fails to successfully deal with the above challenging inputs.

In the end, it shall be stated that the model designed is absolutely real time and accomplishes the 30-s simulation in about 20 s.[xxi] More importantly, never could traditional mean value models or simple fully black-box models reach the accuracy or reliability attained by this gray-box structure. It is worth mentioning that global models normally lack accuracy and local models basically lack wide-range reliability. Nonetheless, here, a global model with good accuracy over a wide range of operation is presented. The model can, in a sense, be regarded as a Hammerstein model followed by another Hammerstein model whose output is fed-back to the system.

2.6 SUMMARY

In this chapter, an accurate real-time engine control-oriented model/simulator (named neuro-MVM) was designed.

As stated, the ordinary models either lack accuracy or fail to reliably account for the entire operating space for all the inputs. One of the approaches in automotive industry is to employ a combination of local models (typically recurrent neural networks), each representing a portion of the engine operating space. We attempted to propose an alternative, which in fact, could be regarded as an extension of mean value models. It incorporated neural networks into subsystems of MVM, such that the merits of each modeling style cover the demerits of the other. The final model is real time with accuracy in wide ranges of engine operation. On top of these, some indispensable subsystems (like knock) vital for a proper simulation, optimization, or control design, were considered in the model.

In neuro-MVM, the system is broken down into simpler subsystem, each of which carefully identified. This decomposition and part-to-whole approach coincides with the cascade nature of the real system as the fluid flow enters subsystems one after the other. The key points regarding the modeling style can be outlined as follows:

- System decomposition to subsystems and making up a combination of black-box identifiers and mathematical relations (or physical insight) to yield a gray-box structure
- Proper design of experiments and data acquisition
- Simplifying the tasks burdened on neural networks, adding assisting knowledge input signals, and local-global modeling, all to improve identification accuracy
- Employing committee structures

As regards data acquisition, different design of experiment approaches could be utilized (such as D-Optimal, or Spherical Screening, etc.). Here, we tried to propose an alternative, specifically designed for this very system.

Ideally, the model should be built based on real experimental data. However, as explained, when enough experimental data is not available, data acquired from a validated CDF model could be assisted.

It shall also be noted that while MVM does suffice for the model-in-the-loop control design, discrete modeling would be required when it comes to hardware-in-the-loop control. Additionally, for more accuracy, actuator and sensor models could also be incorporated. Besides, should the required experimental data be available, a wall-wetting subsystem could also be designed. Wall-wetting in general will be addressed in Chapter 4.

NOTES

i. Note that not all of the papers suffer from all of the problems/challenges.
ii. Note that for this algorithm to work effectively, early stopping needs to be postponed to give the algorithm time to find the best architecture.
iii. This is in fact, the improved version of what is presented in [9] by Brahma, He, and Rutland.
iv. Up to 10 times reduction in MSE, compared to the best individual network trained.
v. Although the current final results, as will be seen, show a high coincidence with the GT-Power results.
vi. We would warmly welcome suggestions from scholars and engineers to improve this approach.

vii. This number is regarded for this case study and could be different depending on the designer's perspective.
viii. Note that the operating intervals (and thus, partitioning) may differ for different cases.
ix. As another example, partitioning for lower throttle angles must be denser than higher throttle angles.
x. Note that in estimation using Taylor series, larger gradients demand smaller gridding, and vice versa.
xi. The outputs of some formulas in gas exchange are indirectly the inputs of the same formulas. Therefore, combining the formulas with a neural network could result in an algebraic loop.
xii. Another input as exhaust manifold pressure could also be added.
xiii. Induction time integral is defined as $T = \int_0^{t_1} dt/\tau \geq 1$, where τ is induction time, which can be calculated from Douaud and Eyzat relation [13].
xvi. due to not having access to the required experimental data
xv. This is in fact the reason why J.S.R. Jang proposed ANFIS in 1993.
xvi. The key idea of improved partitioning was to assign different networks to different output regimes. In this way, the task identification for each network becomes much easier, as each network has to identify one specific regime. In addition, the border area between neighboring regimes is included in the pattern tables of both network structures in order to improve the partitioning quality.
xvii. In fact, most of the dynamic behavior of TWC comes from its oxygen reservoir and becomes important during temporary excursions to lean or rich mixtures due to the wall-wetting dynamic. These temporary excursions can only be well controlled when the after-catalyst oxygen sensor is available.
xviii. As explained in Section 2.3, these results are acquired from an experimentally validated GT-Power software model.
xix. As for HC, due to some technical problems in GT-Power, it is not possible to attain exact *test* results.
xx. Note that the ending 15 seconds of the simulation for catalyst emissions concerns air-fuel ratios below 13, which is highly unlikely for an SI engine. However, the beginning 15s available covers the important prevalent part as to air-fuel ratios between 13–15.
xxi. As mentioned before, the MiL control design contains engine model (designed here) with the ECU model, both simulated in a computer set. HiL encompasses real ECU and real components with real *(or real-time simulated)* process [16]. Interestingly enough since the model here is real time, it could be used for both MiL and HiL steps. The simulation has been carried out with a 3.6-GHz processor with a 16-GB RAM.

REFERENCES

1. Hagan, M.T., H.B. Demuth, and M. Beale, *Neural Network Design*. 1st ed. 1996, Boston, MA: PWS Publishing Company.
2. Haykin, S., *Neural Networks: A Comprehensive Foundation*. 2nd ed. 1999, New Jersey, NJ: Prentice Hall.
3. Mathworks, *MATLAB 2015 Help: How dynamic neural networks work*. 2015.
4. Shamekhi, A.-M. and A.H. Shamekhi, *A new approach in improvement of mean value models for spark ignition engines using neural networks*. Expert Systems with Applications, 2015. **42**: p. 5192–5218.

5. Isermann, R., *Engine Modeling and Control: Modeling and Electronic Management of Internal Combustion Engines*. 1st ed. 2014, Berlin, Germany: Springer.
6. Guzzella, L. and C. Onder, *Introduction to Modeling and Control of Internal Combustion Engine Systems*. 2nd ed. 2010, Berlin, Germany: Springer.
7. Krose, B. and P. van-der-Smagt, *An Introduction to Neural Network*. 8th ed. 1996, Amsterdam, The Netherland: University of Amsterdam.
8. Nelles, O., *Nonlinear System Identification: From Classical Approaches to Neural Networks and Fuzzy Models*. 1st ed. 2000, New York, NY, USA: Springer.
9. Brahma, I., Y. He, and C.J. Rutland, *Improvement of neural network accuracy for engine simulations*. SAE Technical Paper 2003-01-3227, SAE International, 2003.
10. He, Y. and C.J. Rutland, *Application of artificial neural networks in engine modeling*. International Journal of Engine Research, 2004. **5**(4): p. 281–296.
11. Nikzadfar, K. and A.H. Shamekhi, *An extended mean value model (EMVM) for control-oriented modeling of diesel engines transient performance and emissions*. Fuel, 2015. **154**: p. 275–292.
12. Shamekhi, A.-M. and A.H. Shamekhi, *Modeling and simulation of combustion in SI engines via neural networks and investigation of calibration and data acquisition in the GT-power software*. Modares Mechanical Engineering, 2015. **14**(13): p. 233–244.
13. Heywood, J.B., *Internal Combustion Engine Fundamentals*. 1st ed. 1988, New York, NY, USA: McGraw-Hill.
14. Bejan, A., *Heat Transfer*. 1996, New York, NY, USA: John Wiley and Sons.
15. Cortona, E., C.H. Onder, and L. Guzzella, *Engine thermal management with components for fuel consumption reduction*. International Journal of Engine Research, IMechE, 2002. **3**(3): p. 157–170.
16. Isermann, R., J. Schaffnit, and S. Sinsel, *Hardware-in-the-loop simulation for the design and testing of engine-control systems*. Control Engineering Practice, 1999. **7**: p. 643–653.

3 Mid-Level Controller Design

Calibration

3.1 INTRODUCTION

This chapter is to design the engine mid-level controller containing calibration maps. As explained in Chapter 1, due to problem complexity and the conflicting impact of different control inputs on the design objectives, and also the abundance of the engine operating points, manual calibration at a test bench would be highly costly and time-consuming. Therefore, a computer *model-based optimization* is performed.

Table 3.1 outlines some of the major problems/challenges in the literature (in Chapter 1) with their corresponding consequences.[i]

In the light of the arguments made in Chapter 1, in this chapter, based on our methodology, the engine *global* model designed in Chapter 2 is employed. As a matter of fact, this modeling style, for its part, relieves the need for harsh smoothening of calibration maps (as demanded for ordinary *local* models).[ii] Furthermore, the model's accuracy and fast response can enhance the final controllers' quality (and accuracy) and accelerate the optimization procedure, respectively. The model also provides access to intermediate variables, and hence, brings flexibility in optimization.

With regard to optimization, thanks to the model's ability to represent after-catalyst emissions, a constrained driving-cycle-based calibration can be conducted. On top of this, owing to the model's ability to predict knock intensity, safe spark timing calibration could be attained. Moreover, in an effective and simpler approach to filter the variance error resulted from meta-heuristic optimization, the committee method will be employed.

In what follows, after a short glance at the modeling methodology, the optimization strategy will be discussed and subsequently performed [1].

In the end, note that although the air mass flow observer basically belongs to the mid-level layer, for the sake of coherence and better understanding, it will be discussed in Chapter 4.

3.2 A REVIEW ON MERITS OF THE CONTROL-ORIENTED MODELING APPROACH

As previously mentioned, modeling has a profound impact on the calibration quality, even more than the optimization algorithm. As far as optimization is concerned, the model needs to be an accurate and fast response. Inasmuch as

DOI: 10.1201/9781003323044-3

TABLE 3.1
Some of the Major Problems in the Publications Reviewed with Their Consequences, Regarding Model-Based Calibration

Problem/Challenge	Consequences
CFD-model-based calibration	Immensely time-consuming
Local modeling	Perhaps demanding detrimental map smoothening
Not modeling after-catalyst emissions	Not providing a control design based on standard driving cycles
Not modeling knock	Not providing safe operation for a mid-level controller design
Detrimental smoothening approaches for filtering map fluctuations	Sub-optimal control maps

ordinary global models (e.g. mean value modeling [MVM]) lack accuracy, local models are favored in the literature, based on which the subsequent optimization, nevertheless, demands smoothening. This, per se, degrades the accuracy of calibration maps. Accordingly, in this study, a *global* transient model with high accuracy and fast response (designed in Chapter 2) is employed.

Instead of erecting a collection of local models, each representing the entire engine for a limited operation interval, each engine subsystem has been identified separately for the widest range of engine operation, to design a global model.

As explained before, this part-to-whole modeling approach coincides with the system's physical structure, as the working fluid flows from one subsystem to another in series and provides several advantages for engine global modeling. That is, the tasks burdened on identifiers are simplified (as opposed to modeling the entire engine using one identifier), and hence, higher levels of accuracy could be achieved. Furthermore, in this way, some knowledge regarding the system's physical structure is incorporated into the model, making the resulting gray-box structure more reliable than a mere black-box model. What is more, this approach provides access to intermediate variables and provides flexibility in optimization.

On top of these, the model (re-displayed in Figure 3.1) includes all the necessary inputs and outputs for an appropriate optimization. Especially, the model is able to represent the engine's after-catalyst emissions and knock intensity, which are of great use in the calibration phase.

3.3 MODEL-BASED OPTIMIZATION

The task of the mid-level layer in a hierarchical torque-based control structure is to calculate optimal values for engine actuators, at each operating point. In other words, the mid-level controller is expected to present optimal values for actuators at the instantaneous engine speed n_i, ambient conditions, and the desired torque T_i (or desired air mass flow rate $\dot{m}_i$) demanded from the high-level controller. The optimization is conventionally performed for steady-state conditions. This will be further discussed later.

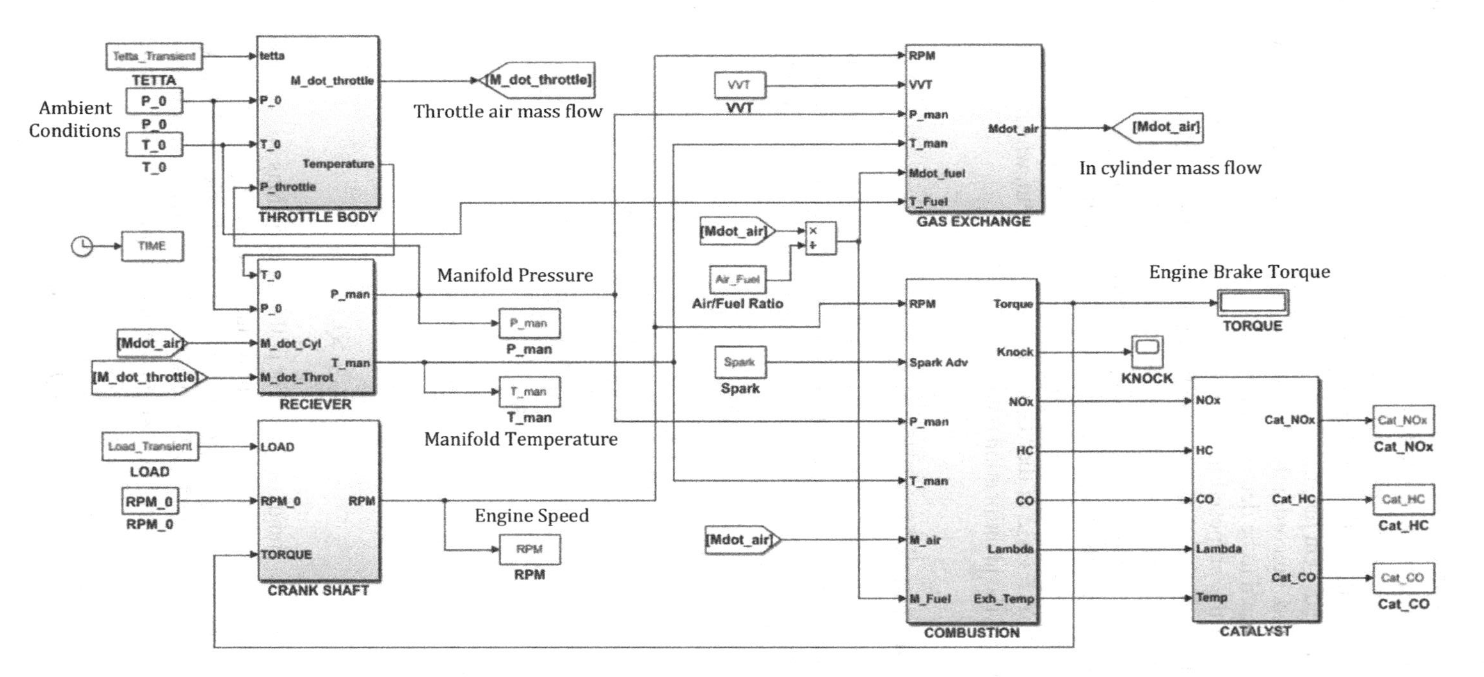

FIGURE 3.1 The final model architecture. Re-display of Figure 2.32.

As explained earlier, in general, there are two main engine operating modes: part-load operation mode (normal mode) and full-load operation mode.[iii] With regard to normal mode, the most efficient engine operation within emission limits is sought, that is, the optimal values of throttle angle, VVT, and spark advance need to be derived.[iv] On the other hand, as for full-load mode, merely, maximum torque is invoked. Accordingly, the optimal values of air-fuel ratio, VVT, and spark advance need to be determined for full-load mode.

A benefit of the proposed model (Neuro-MVM) is that it provides access to intermediate parameters and provides flexibility in optimization, as each subsystem has been identified separately. For steady-state conditions, there would be no difference whether the optimization is performed for each specific speed-torque (n_i, T_i) or speed-throttle angle (n_i, θ_i), as both cases could be easily converted to each other.[v] Nevertheless, for some torque-based strategies, it might be required to build the (off-line) inverse of the "Combustion" subsystem, which could be easily carried out with the identification data in hand.

3.3.1 Optimization Methodology

In the normal mode, the optimization target is to minimize brake-specific fuel consumption (BSFC), under the constraints specified for pollutant emissions and knock intensity. That is, for each operating point (n_i, T_i), or equivalently (n_i, θ_i), and in the ambient conditions (P_0, ϑ_0):

$$\min_{\mathrm{u}} \mathrm{BSFC}\left(u, n_i, \theta_i\right)$$

s.t.:

$$\bar{m}_{CO} \le m_{CO,\lim}$$

$$\bar{m}_{HC} \le m_{HC,\lim}$$

$$\bar{m}_{NOx} \le m_{NOx,\lim}$$

and[vi]:

$$\mathrm{Index}_{\mathrm{Knock}} \le \mathrm{Lim}_{\mathrm{Knock}}. \tag{3.1}$$

However, commonly, standard limits for pollutant emissions ($m_{CO,\lim}$, $m_{HC,\lim}$, and $m_{NOx,\lim}$) are not specified for every individual operating point, but rather, over standard driving cycles (e.g. New European Driving Cycle [NEDC], Worldwide harmonized Light vehicles Test Procedure [WLTP], etc.). Accordingly, a multi-variable objective function with weighting factors needs to be formulated [2]. That is

$$\min_{\mathrm{u}} J\left(u, n_i, \theta_i\right)$$

s.t.:

$$\text{Index}_{\text{Knock}} \leq \text{Lim}_{\text{Knock}}$$

$$\bar{m}_{\text{CO,DC}} \leq m_{CO,\lim}$$

$$\bar{m}_{HC,DC} \leq m_{HC,\lim}$$

$$\bar{m}_{NOx,DC} \leq m_{NOx,\lim}$$

and

$$J\left(u,n_i,\theta_i\right)=w_1\left(\text{BSFC}\Big/\overline{\text{BSFC}}\right)+w_2\left(NOx\Big/\overline{NOx}\right)+w_3\left(CO\Big/\overline{CO}\right)+w_4\left(HC\Big/\overline{HC}\right) \tag{3.2}$$

in which, $\bar{x}$ denotes the normalization factor for x, in g/kW-h, $\bar{m}_{y,DC}$ is the amount of pollutant y emitted over the course of the desired driving cycle (DC), in gram, w_i are the weighting factors, whose sum equals to 1. That is

$$\bar{m}_{y,\text{DC}}=\oint_{\text{DC}} dm_y, \quad \text{and} \quad \sum_{i=1}^{4} w_i = 1$$

$\text{Lim}_{\text{Knock}}$ can be set to 200 [3]. To accelerate the optimization process, the knock limit constraint could be incorporated into the function as an extra term:

$$\acute{J}\left(u,n_i,\theta_i\right)=J\left(u,n_i,\theta_i\right)+\acute{w}\left(\text{Index}_{\text{Knock}}\Big/\overline{\text{Index}_{\text{Knock}}}\right) \tag{3.3}$$

where

$$\text{Index}_{\text{Knock}}=\begin{cases} 0, & \& \ \text{Index}_{\text{Knock}} \leq \text{Lim}_{\text{Knock}} \\ \text{Index}_{\text{Knock}}, & \text{Index}_{\text{Knock}} > \text{Lim}_{\text{Knock}} \end{cases}.$$

Other constraints should also be incorporated to avoid impermissible values for BSFC, spark advance, VVT, etc.

The multi-variable weighted function J is derived based on the driving cycle[vii] (i.e. by meeting the driving cycle standard limits) and then is applied *to calibrate the whole engine operating space*. We call it *driving-cycle-based (or cycle-based calibration).*

With regard to full-load mode, the target is to maximize the brake torque. In these extreme circumstances, emission standard limits are ignored

$$\max_{\text{u}} T\left(u,n_i\right)$$

s.t.:

$$\text{Index}_{\text{Knock}} \leq \text{Lim}_{\text{Knock}}$$

where

$$T\left(u,n_i\right) = \text{Brake Torque for full} - \text{load operation.} \tag{3.4}$$

As explained before, since the objective functions are mostly susceptible to be non-convex, meta-heuristic algorithms are employed. Nevertheless, these algorithms do not yield accurate results, but rather, approximate ones. Here, to resolve the consequent fluctuations, ensemble averaging of network committees will be utilized.[viii] Most famous of the committee methods, it has an effective capability of dampening variance error and overcoming bias/variance dilemma [4]. In this way, smooth maps are achieved without any deviation of the results from their optimal values, and merely the variance error is filtered.[ix] Figure 3.2 portrays a basic schematic of ensemble averaging.

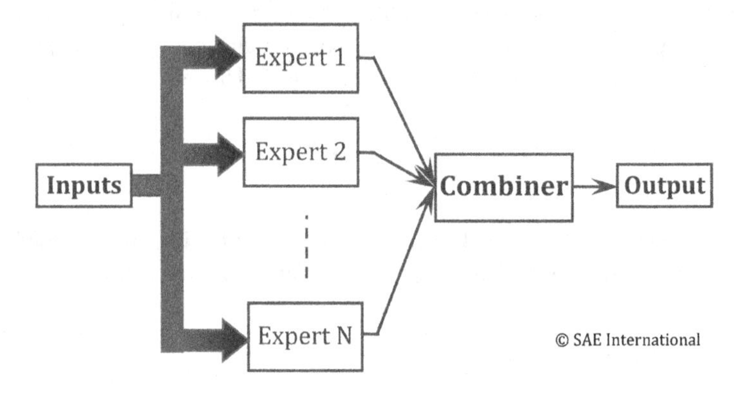

FIGURE 3.2 A basic schematic of ensemble averaging [1].

The octane number for the fuel employed is 94. It should be noted that VVT is usually optimized independently of other variables and is merely based on maximizing in-cylinder mass flow rate. This will be addressed in Section 3.3.3, preliminary optimization.

The final results could (later, in the SiL stage) be stored in multiplicative 2-D look-up tables (or also tables with nested structure) [2], or in LOLIMOT, LOPOMOT, or adaptive polynomial structures [5, 6]. Anyhow, the employed structure needs to be appropriate for an engine ECU, in terms of floating-point operations per second (FLOPS),[x] required memory, and of course, accuracy.

Figure 3.3 depicts a simple example of a multiplicative look-up table structure. Since look-up tables of dimension above 2 are inefficient, they are broken down into several multiplicative look-up tables. As can be seen, the two most influential (primary) inputs form the main look-up table, which is then modified by a correcting factor designed based on a less influential input. Note that the correcting factor could also be a 2-D look-up table, as well. Plus, there could be several look-up tables multiplying (or adding) with each other.

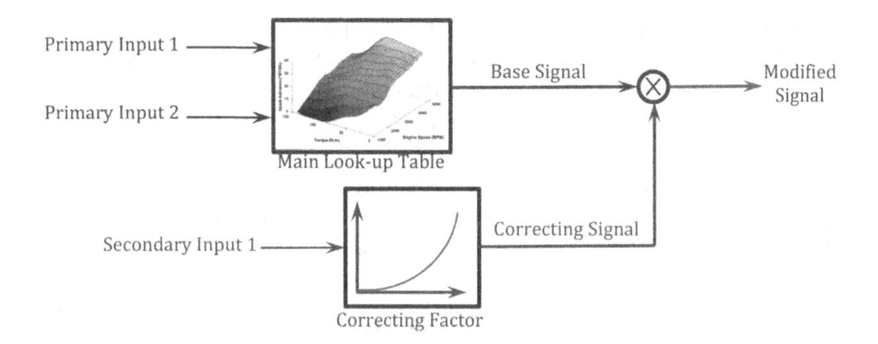

FIGURE 3.3 A simple example of a multiplicative look-up table.

In what follows, after a short glance at the existing optimization algorithms, a preliminary optimization is performed. Afterward, the main optimization will be conducted. In the end, full-load mode calibration will be examined.

3.3.2 A Review of Optimization Algorithm Selection

Due to catalyst and the conflicting behavior of BSFC, emissions, and knock, the optimization problem is prone to be non-convex, and global meta-heuristic methods need to be employed. Several factors play parts in the efficacy of an optimization algorithm, such as diversification, intensification, run-time, etc. A very short review of some optimization algorithms is provided in Appendix B.

Civicioglu and Besdok, in a comprehensive study, compared cuckoo-search (CK), particle swarm optimization (PSO), differential evolution (DE), and artificial bee colony (ABC) algorithms by testing over 50 different famous benchmark functions. According to their study, the DE algorithm performs better than the other algorithms, in terms of run-time complexity and the required function-evaluation number for acquiring global minimizer. Furthermore, the results of DE and CK algorithms are more precise and robust compared to the ABC and PSO algorithms [7]. Kachitvichyanukul compared the three most popular methods GA, PSO, and DE. It is demonstrated that PSO, due to its high intensification, and yet, low diversification is susceptible to pre-mature convergence [8]. In another work, Lim and Haron compared the above three algorithms for the same parameters, setting, and optimization problems. They indicated that GA can attain highest number of best minimum fitness values and converge faster compared to DE and PSO [9].

In the light of the above arguments, genetic algorithm could still be considered as one of the best optimization algorithms, with appropriate diversification and intensification and is applied in this study.

Since meta-heuristic optimization is rather slow, the algorithm parameters should be carefully adjusted to reach enough accuracy within least computational time necessary. As for GA, the required solution time rises exponentially with population size. Furthermore, in MATLAB, too large values for "MaxStallGenerations" parameter or too small values for "FunctionTolerance" parameter would lead to fruitless over-computation.

As regards independently calibrating VVT (Section 3.3.3), since the optimization problem is convex, local gradient-based algorithms can be employed. Not only do these methods present accurate solutions (providing that the problem is convex), but they also respond much faster than global algorithms. Here, "fmincon" MATLAB function, which is based on "interior-point" method, is employed.

3.3.3 First Step: Preliminary Optimization

In this step, VVT is first calibrated to yield maximum in-cylinder air mass flow rate[xi] at each operating point (n_i, θ_i, P_0, and ϑ_0). The VVT for the engine in hand is of the variable camshaft timing (VCT) type. As explained, being a convex problem, VVT optimization could be carried out by "fmincon" MATLAB function. The model required for this procedure includes the combination of "Throttle body," "Receiver," and "Gas exchange" subsystems in Figure 3.1. The optimization is conducted for 2834 different cases of (n_i, θ_i, P_0, ϑ_0). The VVT surface for standard atmospheric ambient conditions is represented in Figure 3.4. The distortion in the surface conforms to the distortion in typical engine volumetric efficiency curves. Note that the optimization has been solved without any constraints. In practice, there may be limits for varying camshaft timing. Accordingly, there could be upper and lower saturation bounds, for instance, as $217° \leq VVT \leq 242°$ (with respect to firing-TDC).

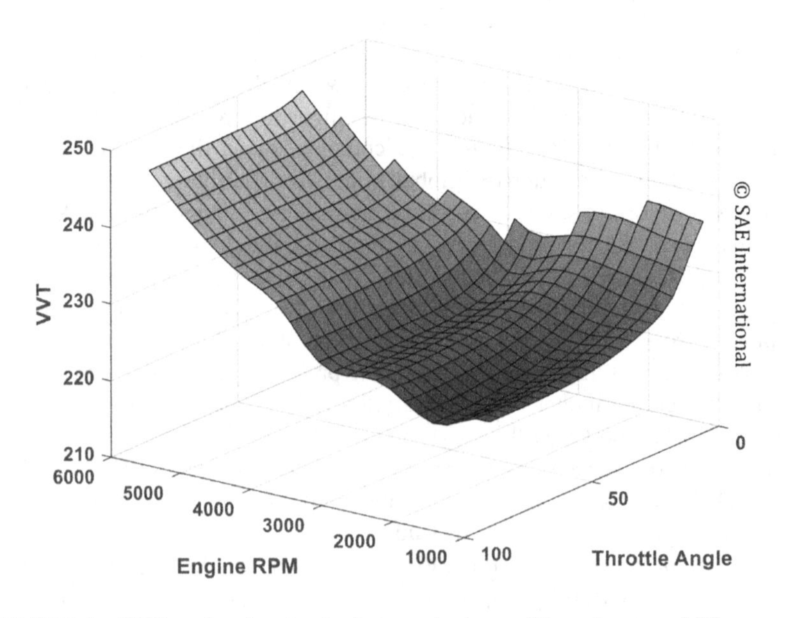

FIGURE 3.4 VVT surface for standard atmospheric conditions, in general [1].

Next, based on (some points of) an already known calibration map, a weighted objective function, including terms BSFC, knock, and engine-out emissions (in ppm, without catalyst) is derived:

$$F = \frac{\text{BSFC}}{100} + \frac{\text{NO}x}{5000} + \frac{\text{CO}}{12{,}000} + \frac{\text{HC}}{3500} + \frac{\text{Knock}}{2000}. \quad (3.5)$$

For this stage, the "Combustion" subsystem can be utilized as the model. The resulting values for standard atmospheric ambient conditions could be seen in Figure 3.5. Also, the calibration map for throttle angle is represented in Figure 3.6.

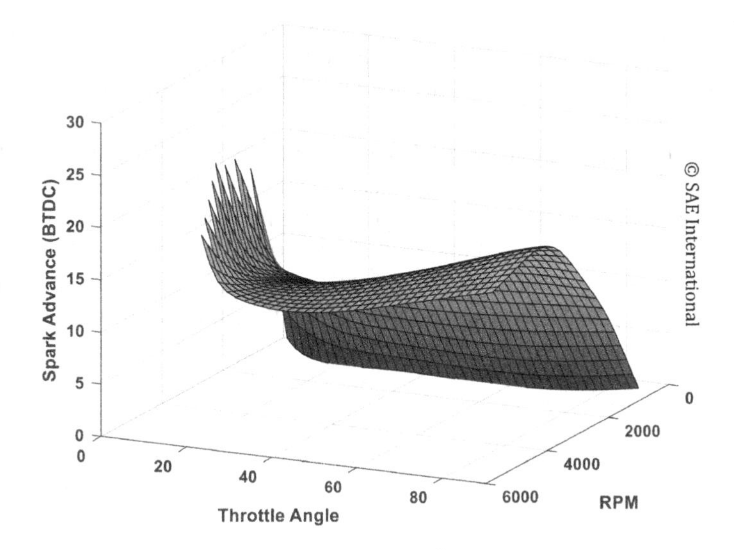

FIGURE 3.5 Preliminary calibration map for spark advance [1].

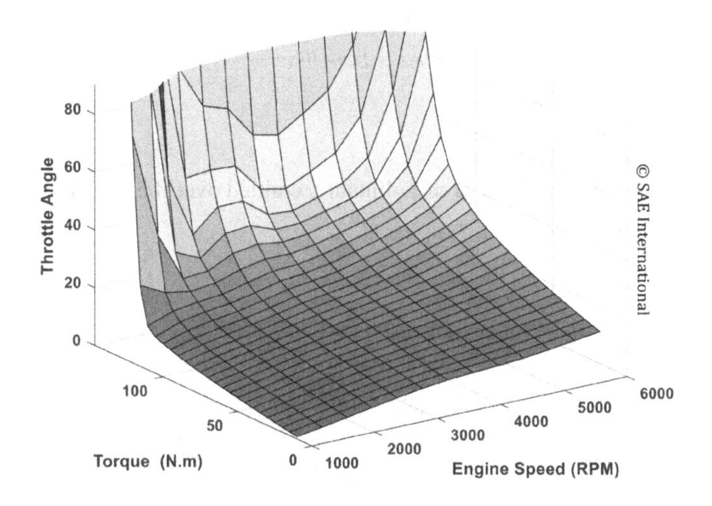

FIGURE 3.6 Preliminary throttle angle map [1].

3.3.4 Second Step: Main Calibration

In the light of the previously mentioned arguments, in this step, first a weighted objective function using a standard driving cycle needs to be derived, based on which, the whole engine operating space is then calibrated. Since the engine in hand is designed and produced based on the NEDC, this very standard cycle is employed for this case study. Needless to say, the methodology presented here is independent of driving cycle, and could be easily utilized for other driving cycles.

3.3.4.1 Determination of the Weighted Objective Function

NEDC consists of specific operating points with specified speeds and gear changes, and thus, vehicle longitudinal dynamic needs to be taken into account. The goal is to determine the weights in equation (3.2) and formulate the objective function $\acute{J}$, equation (3.3).

Hence, the optimization problem (3.6) given below must be solved for NEDC points in an iterative procedure. After each iteration, using the derived optimal control variables $\boldsymbol{u}$, the sum of each pollutant emission "y" over the course of NEDC is calculated ($\bar{m}_{y,\mathrm{NEDC}}$) and compared with the respective standard limit ($m_{y,\lim}$). If the limits are not met, the weighting factors, w_i, are modified, and this procedure is reiterated over and over, until finally the emission standard limits are met. Figure 3.7 represents the pertinent flowchart.

$$\min_{\mathrm{u}} \acute{J}\left(u, n_i, T_i\right)$$

s.t.:

$$\bar{m}_{CO,\mathrm{NEDC}} \le m_{CO,\lim}$$

$$\bar{m}_{HC,\mathrm{NEDC}} \le m_{HC,\lim}$$

$$\bar{m}_{NOx,\mathrm{NEDC}} \le m_{NOx,\lim}$$

where

$$\bar{m}_{y,\mathrm{NEDC}} = \int_{\mathrm{NEDC}} dm_y = \text{The sum of pollutant } y \text{ emitted over the course of NEDC.} \tag{3.6}$$

The following relation can be utilized for weight adaptation (inspired from [10]):

$$w_y^{k+1} = w_y^k \left(1 + \frac{\bar{m}_{y,\mathrm{NEDC}} - m_{y,\lim}}{m_{y,\lim}}\right)$$

$$w_{\mathrm{BSFC}}^{k+1} = 1 - \sum_y w_y^{k+1}. \tag{3.7}$$

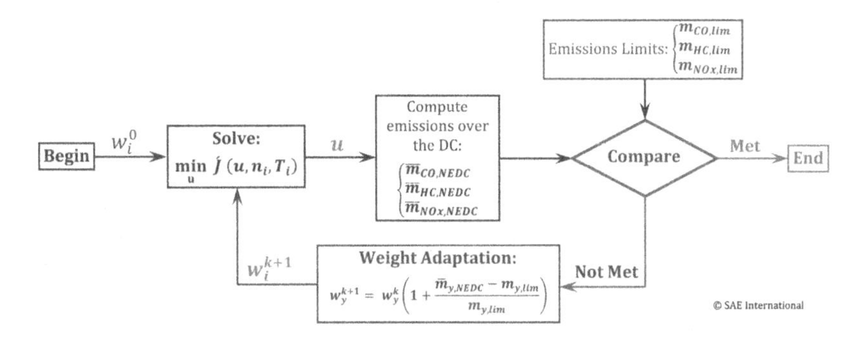

FIGURE 3.7 The flowchart for determining the weighed objective function [1].

Thanks to the model's ability to represent transient behavior, the optimization in each iteration could be performed for the whole NED cycle (with its transients), and thus, a dynamic calibration is accomplished.[xii] Nevertheless, this entails huge computational burden and demands extremely powerful computers. A simpler approach is to conduct the optimization for a number of average points as representatives of the driving cycle (inspired from [2]). As mentioned earlier, this necessitates considering the vehicle's longitudinal dynamic. Table 3.2 presents the longitudinal specifications of the vehicle. Figure 3.8 displays 17 most occurring points, as representatives of NEDC for the vehicle. The relative contribution of each point in the driving cycle is represented by the size of the relevant circle. Note that the locations of the points depend upon the vehicle specifications and differ from vehicle to vehicle. Obviously, for more accurate optimizations, a greater number of representatives could be chosen.

TABLE 3.2
The Vehicle's Longitudinal Specifications

Specification	Value
Weight	1200 kg
Length	4502 mm
Width	1720 mm
Height	1460 mm
Tire	185/65 R15
1st Gear Ratio	3.45
2nd Gear Ratio	1.86
3rd Gear Ratio	1.29
4th Gear Ratio	0.95
5th Gear Ratio	0.75
Final Drive Ratio	4.53

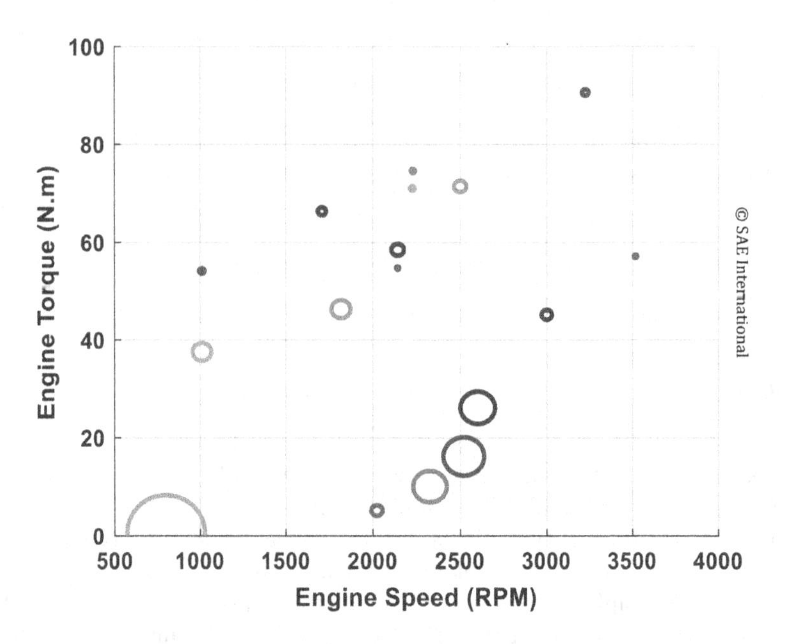

FIGURE 3.8 The 17 average points as representatives of NEDC, as an example (inspired from [2]). The relative contribution of each point in NEDC is indicated by the size of the relevant circle [1].

As explained, in each iteration, control variables $\boldsymbol{u}$ are optimized for the representatives.[xiii] Afterward, the sum of each pollutant emission "y" over the course of the driving cycle must be calculated $\left(\bar{m}_{y,\text{NEDC}}\right)$ and compared with the standard limit to update the weights using relation (3.7). However, the control variables are only derived for the representatives, not the whole cycle, and thus, the parameter $\bar{m}_{y,\text{NEDC}}$ has to be estimated based on sum of the representatives of the cycle (with their relative contribution).

Now, at this stage, for better accuracy, the whole engine operating space could be optimized using the derived objective function above and sum of each pollutant emission "y" over the course of the driving cycle could be re-calculated and compared with the standard limit. Subsequently, the respective weight w_y can be adapted again, using relation (3.7), to obtain a more accurate updated objective function, before moving to the next iteration.

The whole procedure should be re-iterated over and over, until the standard limits are met. The final attained objective function (as mentioned) is employed to calibrate the entire operating space. Figure 3.9 indicates the convergence of w_1, as an instance, for the NOx emission.[xiv]

The standard limits could be regarded stricter in order to upgrade the final design to a higher standard. It shall be reiterated that the methodology for other standards and driving cycles would be the same and could be easily generalized.

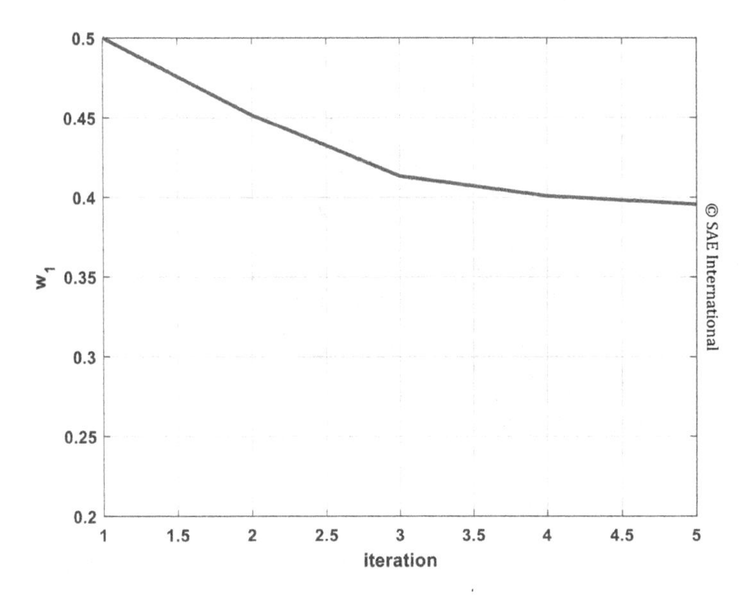

FIGURE 3.9 The convergence of w_1, as an example [1].

3.4 FINAL RESULTS

3.4.1 Calibration Results: Normal Mode

Based on the weighted objective function derived in the previous section, the entire engine operating space is calibrated. A number of 782 steady-state operating points with different (n_i, T_i, P_0, and ϑ_0) are optimized, and all the respective optimal variables are acquired. As stated before, the results need to be first taught to network committees (for variance error elimination)[xv] and then could be stored in the ECU as multiplicative or nested look-up tables or some efficient local-model-tree structures such as LOILIMOT or LOPOMOT networks.

Figures 3.10–3.12 represent the calibration maps of spark advance, throttle angle, and VVT for standard atmospheric ambient conditions, with respect to the demanded torque and engine speed (for this case study). It shall be noted that the calibration has been conducted for different ambient conditions, but due to restrictions in 3-D representation, they are only depicted for standard atmospheric ambient conditions. Figure 3.13, also, displays the final BSFC contours for standard ambient conditions. The results are clearly much less conservative compared to the preliminary calibration maps. More importantly, the resulting maps are completely smooth, thanks to the methods adopted i.e. global modeling and ensemble averaging.

As expected, when the engine speed rises, ignition timing must be advanced (Figure 3.10). This is owing to the fact that, even though turbulent flame speed increases proportionally with engine speed, the ignition delay (i.e. the time delay

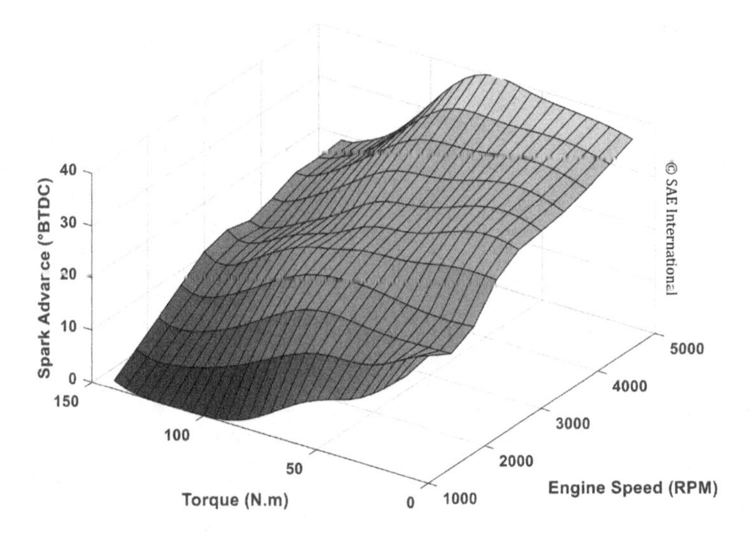

FIGURE 3.10 Spark advance calibration map for standard ambient conditions [1].

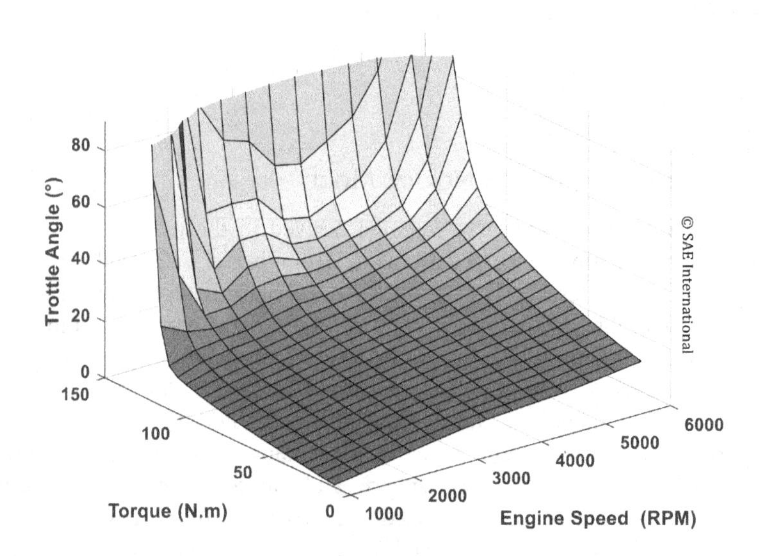

FIGURE 3.11 Throttle angle map for standard ambient conditions [1].

for ignition initiation) remains almost constant. Accordingly, the corresponding crank angle delay rises as engine speed increases.

For smaller torque demands, smaller throttle angles are required (Figure 3.11) to produce less energy. This leads to larger spark advances (Figure 3.10). This is simply because turbulent flame speed declines in the resulting lower cylinder pressures, demanding more time (and thus, more crank angle interval) to complete combustion.

Also, as expected, valve timing shall be raised at higher engine speeds (Figure 3.12), for the sake of ram effects [12]. Also, BSFC contours are correctly of onion shape, with best engine efficiency around 120 N·m and 2750 rpm (for this case study).

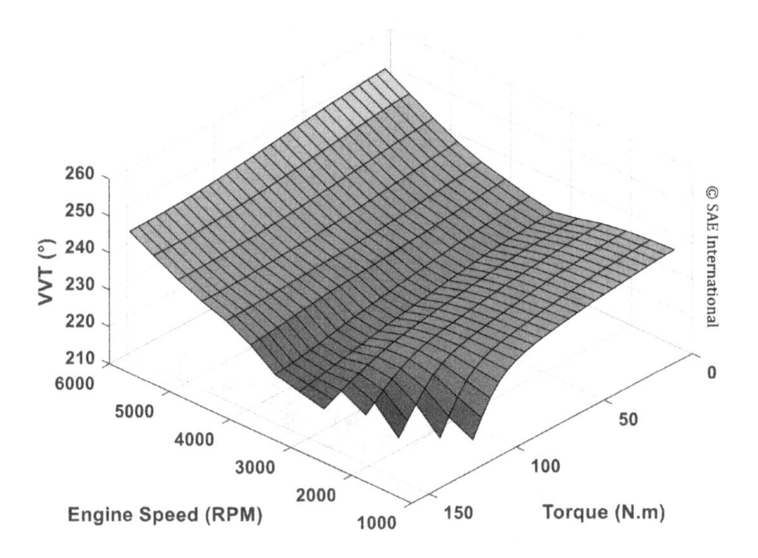

FIGURE 3.12 VVT calibration map for standard ambient conditions, in general. The saturation limits is 217° ≤ VVT ≤ 242° [1].

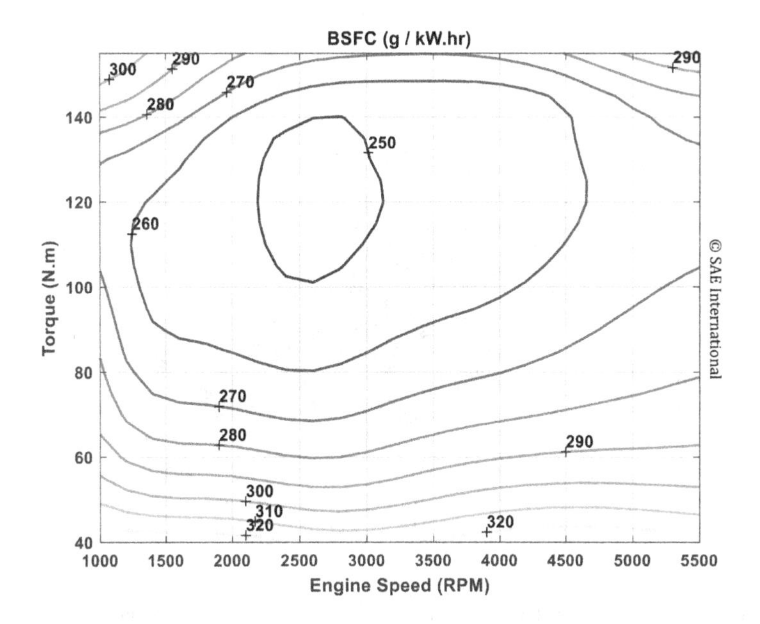

FIGURE 3.13 BSFC contours for standard ambient conditions [1].

The calibration maps, instead of desired torque, could also be expressed with respect to desired air mass flow rate (or desired fuel mass flow rate) into cylinders. Figures 3.14 and 3.15 indicate the effects of different ambient conditions and desired air mass flow rates on throttle angle for 4000 rpm. As can be seen, the

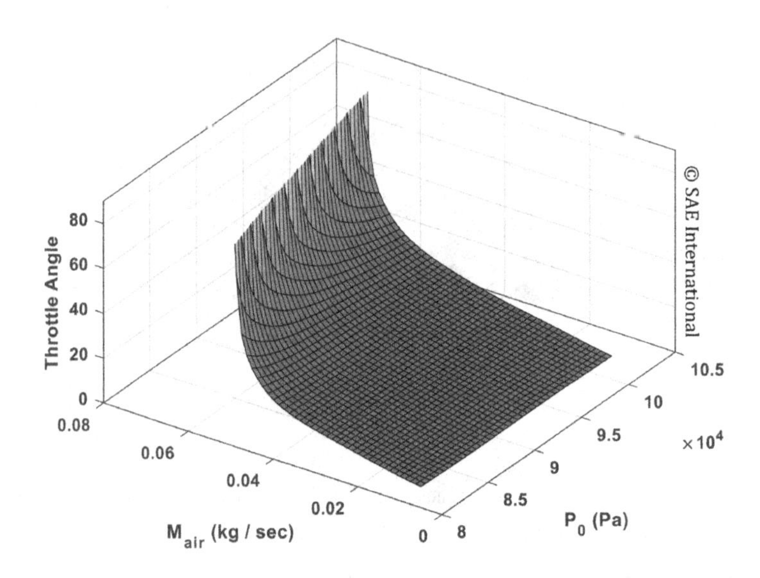

FIGURE 3.14 The impact of desired air mass flow rate and ambient pressure on throttle angle for 4000 rpm [1].

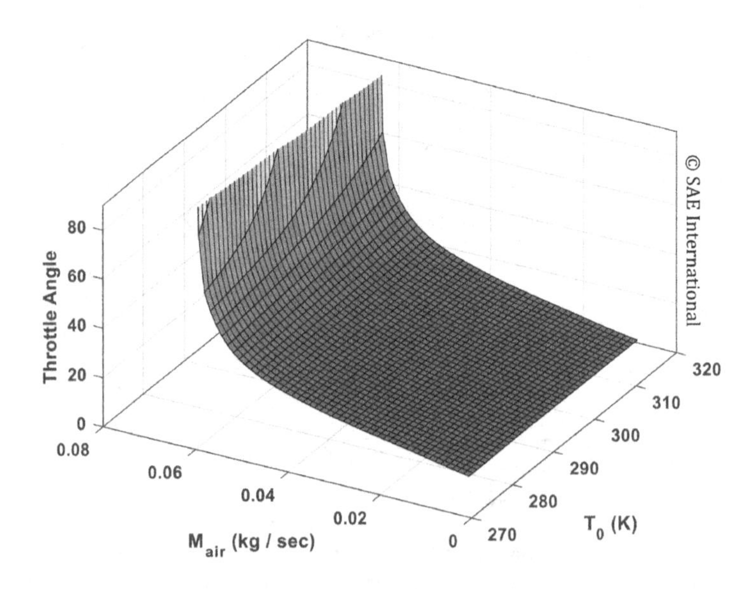

FIGURE 3.15 The impact of desired air mass flow rate and ambient temperature on throttle angle for 4000 rpm [1].

impacts of ambient conditions, particularly as for ambient temperature, are small, uniform, and fairly linear. Accordingly, they could be considered as correcting factors in multiplicative look-up tables, modifying the base signal from the main look-up table. The primary inputs would be the engine speed and the desired air mass flow rate into cylinders.

3.4.2 Calibration Results: Full-Load Mode

In wide-open throttle (full-load mode), the optimization relation (3.4) is solved for various engine speeds and ambient conditions. Figures 3.16 and 3.17 represent the resulting spark advance and air-fuel ratio for different speeds and ambient conditions. As can be seen, the effect of ambient pressure and temperature on control variables are not intricate, and hence, multiplicative look-up tables could be easily used for storage in ECU.

Just like part-load, spark advance rises with increase in engine speed (Figure 3.16). What is more, according to Figure 3.17, maximum torque for full-load mode generally occurs at air-fuel ratios within 11.5–12.5.

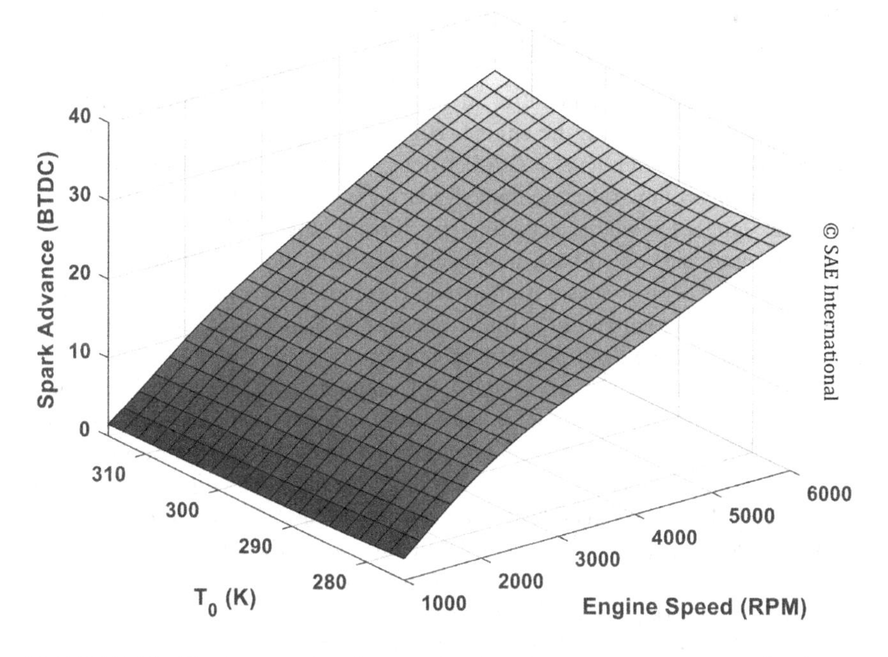

FIGURE 3.16 The impact of engine speed and ambient temperature on spark advance (BTDC) [1].

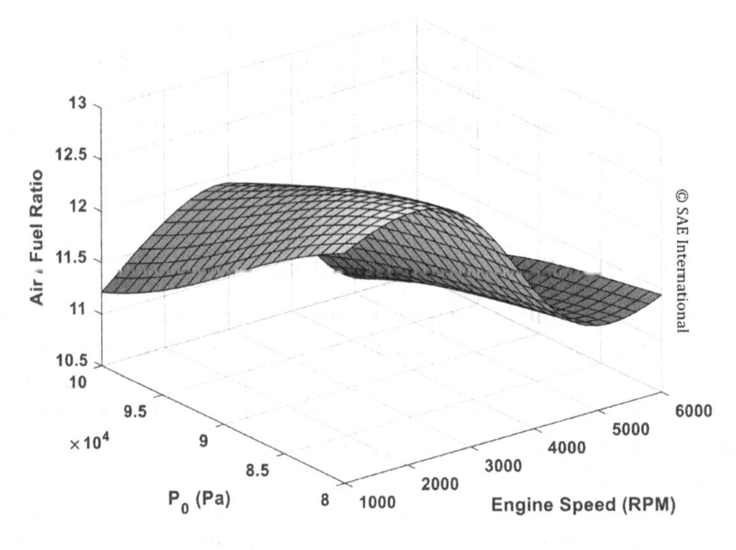

FIGURE 3.17 The impact of engine speed and ambient pressure on air-fuel ratio [1].

3.5 FURTHER DISCUSSION REGARDING THE FINAL STRUCTURE

Figure 3.18 depicts the final control structure, including feed-forward (FF) mid-level controllers regulating the engine. High-level and low-level controllers are not displayed. The high-level controller determines the required torque (or the required air mass flow rate into the cylinder). FF Mid-Level controllers provide base-line signals for engine actuators (i.e. throttle angle, VVT, spark advance, and fuel command). Feed-back low-level controllers modify the base-line fuel command (based on λ-sensor signal) and the base-line spark advance (based on knock sensor signal and in-cylinder pressure or ion-current sensor signals, if available), making the control system more robust.

As can be seen, throttle angle and VVT commands are directly determined based on the instantaneous engine speed, desired torque (or desired air mass flow rate, from high-level layer), and of course, ambient conditions. However, the FF fuel command is determined based on the *instantaneous air mass flow rate* into the cylinder (rather than desired air mass flow rate). Should Mass Air Flow (MAF) sensor (which measures the instantaneous air mass flow rate) not be available, an MAF observer is incorporated in order to estimate the *instantaneous* air mass flow rate into cylinders (Sections 1.4.2.2 and 4.2).

An interesting advantage of designing the maps based on air mass flow rate (instead of the desired torque) is that spark advance controller (and consequently the whole mid-level control structure), besides steady states, could also successfully account for transients.[xvi] Had spark advance been regulated based

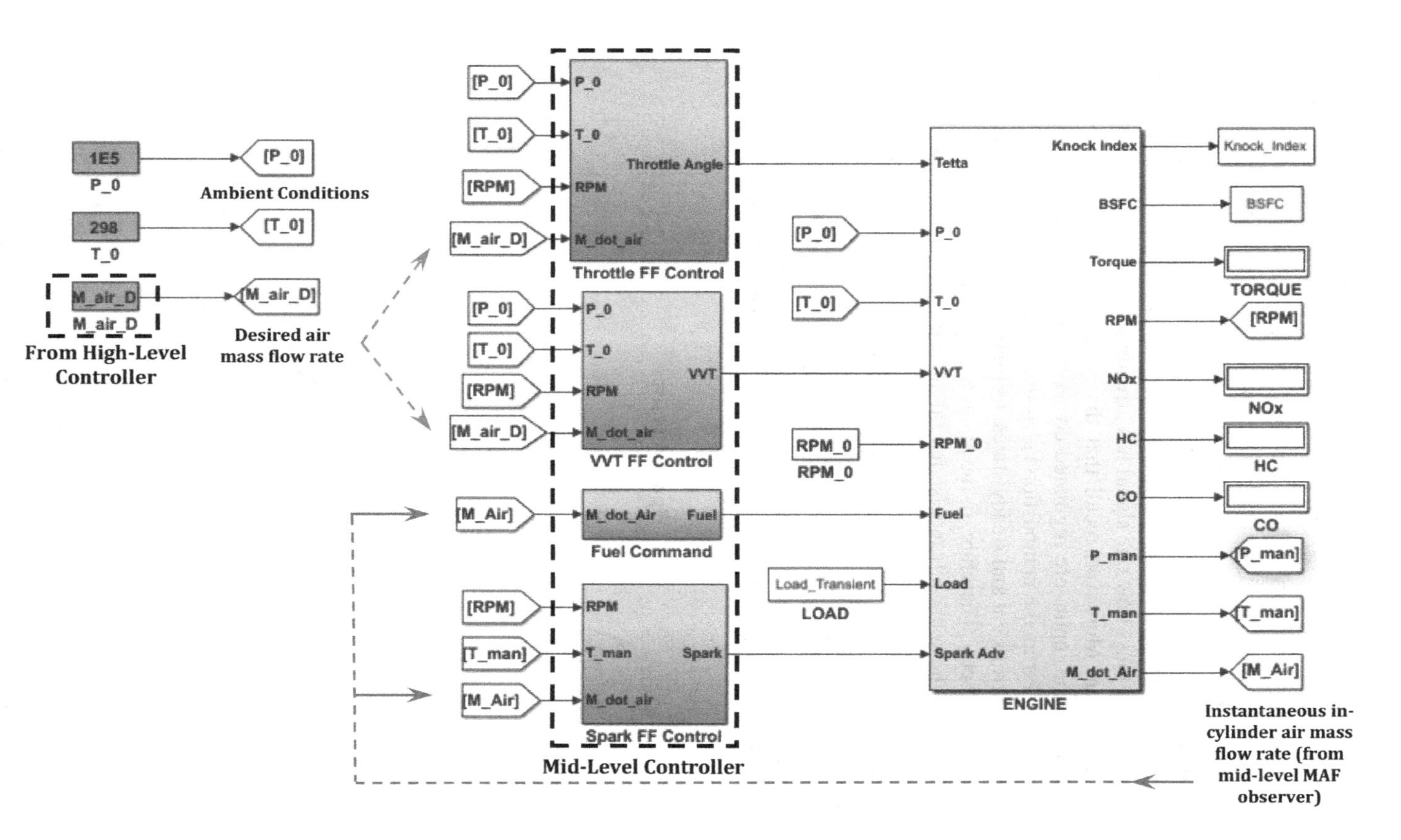

FIGURE 3.18 The final MiL combination of the mid-level controller and the engine model. The MAF observer is not shown.

on the desired torque, manifold dynamic would not have been taken into consideration for combustion control. In other words, in this flow-rate-based control, *a sort of dynamic calibration is achieved.* This means that, although the calibration maps in hand were derived from steady-state optimization, they could also account for a transient-state behavior. To explain more, note that the *instantaneous* air mass flow rate signal (from the MAF observer in the mid-level layer or the modified fuel signal in the low-level layer) is in fact downstream of the high-level layer (with desired torque signal in the former structure). Furthermore, this instantaneous air mass flow rate signal has already taken the manifold dynamic into consideration. Moreover, recall that the combustion subsystem is static, and thus, once the inputs are regulated correctly, the outputs will be proper; that is to say, as far as the combustion subsystem is concerned, there is no difference between transient and steady states, as long as the *instantaneous* inputs are adjusted correctly. Therefore, in the air-flow-based control structure, during transients, since the spark advance signal (along with the fuel injection signal) is adjusted based on the instantaneous air mass flow rate into cylinders, the calibration maps in hand are appropriate although they are derived based on steady-state calibration.

In the end, examples of the control input parameters for the major part of NEDC (with an ordinary first-order high-level controller) are depicted in Figures 3.19–3.21. The resulting engine speed is displayed in Figure 3.22.

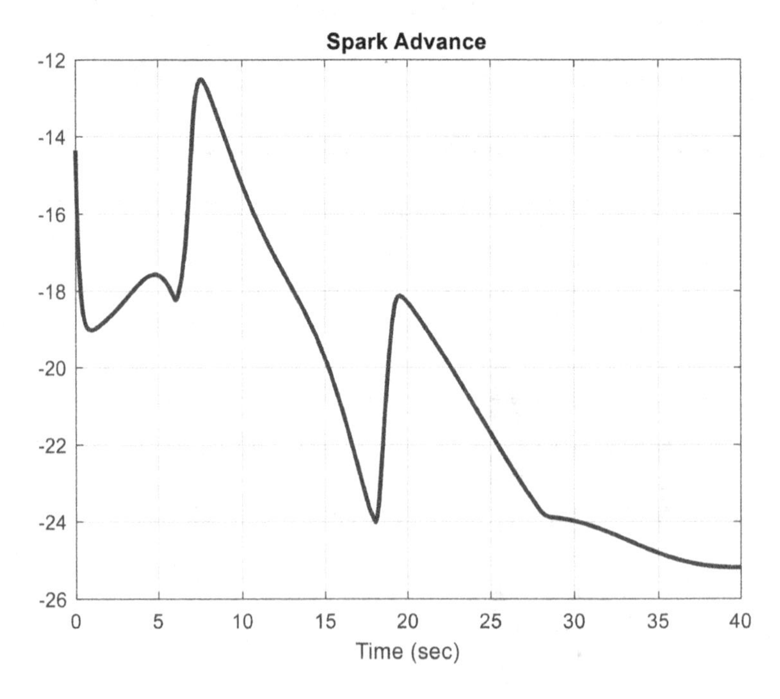

FIGURE 3.19 The resulting spark advance command for the main part of NEDC.

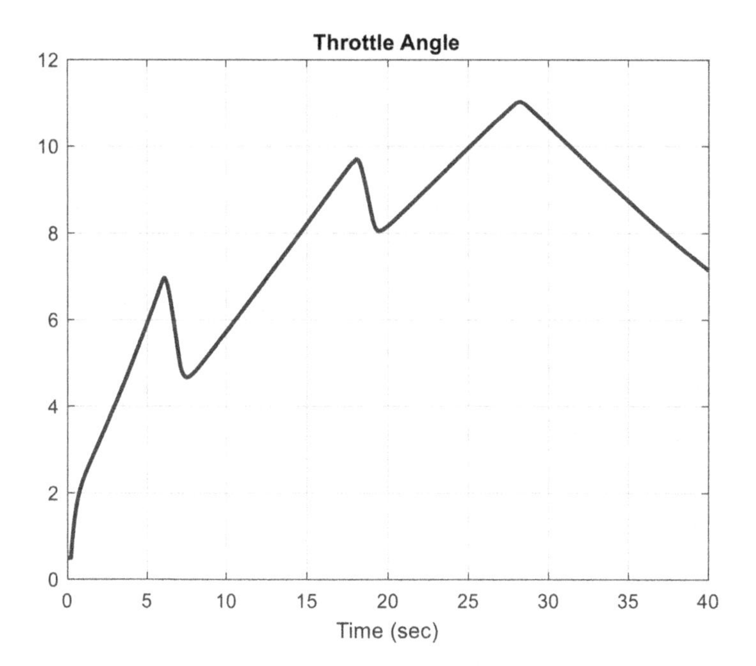

FIGURE 3.20 The resulting throttle angle command for the main part of NEDC.

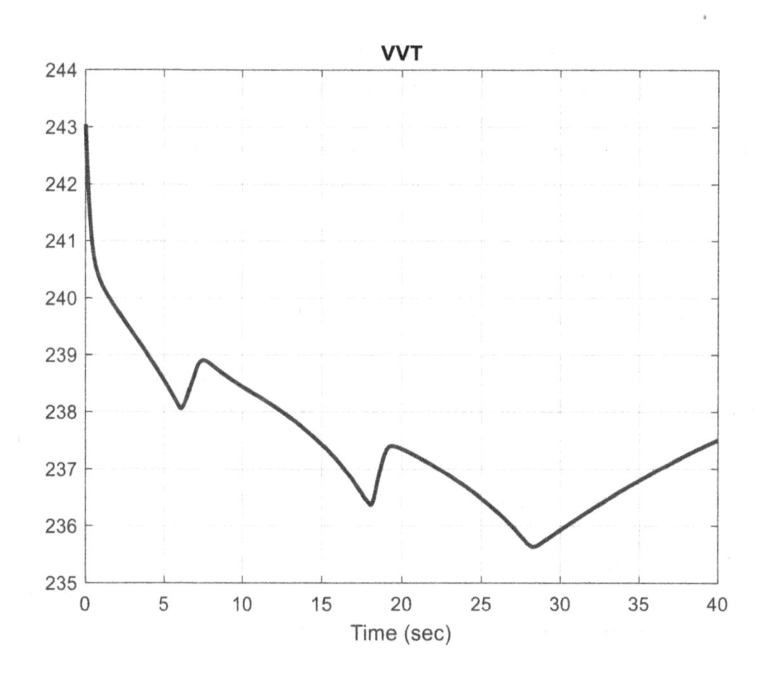

FIGURE 3.21 The resulting VVT command for the main part of NEDC.

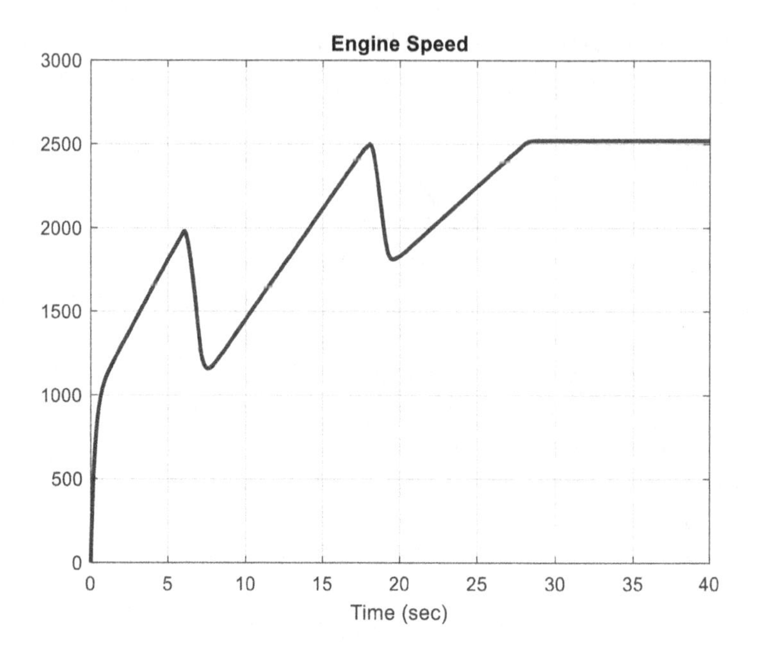

FIGURE 3.22 The resulting engine speed for the main part of NEDC.

3.6 SUMMARY

In this chapter, model-based calibration of SI engines was examined, and a methodology based on standard driving cycles was presented.

In terms of modeling, instead of local modeling or CFD modeling, the engine global control-oriented model designed in Chapter 2 was employed for calibration. As explained, the model's accuracy and fast response could enhance the final controller quality and accelerate the optimization process, respectively. Additionally, this global modeling style, for its part, could relieve the need for detrimental smoothening of calibration maps as demanded for ordinary local models. The model also provides access to intermediate variables, and thus, brings flexibility in optimization.

As regards optimization, a constrained driving-cycle-based calibration was performed. As noted, with the vehicle's longitudinal dynamic taken into account, after-catalyst emissions over the course of driving cycle and engine knock were regarded as constraints in the optimization. A weighted multi-variable objective function based on the driving-cycle's emission limits was derived in an iterative procedure, based on which the whole engine operating space was calibrated. The engine operation was divided into two modes of part-load and full-load, and relevant calibration maps were derived for different ambient conditions. Moreover, in an effective and simple approach to relieve variance errors resulted from meta-heuristic optimizations, the committee method was employed and smooth maps were achieved, without any deviation of the results from optimal values.

In the end, mid-level controller calibration maps for throttle angle, VVT, spark advance, and air-fuel ratio (for full-load mode) were obtained in widest ranges of engine operation.

Note that the methodology presented here could be generalized to any driving cycle and standard. Moreover, for more reliability, exhaust gas temperature could also be considered as a constraint.

NOTES

i. Not all the papers reviewed suffer from all the problems stated.
ii. It shall be reiterated that switching from one local model to another local model could make the results uneven, and thus, necessitate smoothening. This smoothening deviates the calibration map from the derived optimal values. Of course, applying meta-heuristic optimization algorithms could also bring about uneven maps and worsen the problem.
iii. Additionally, depending on the designer's perspective, idle and cold start could also be separated as distinct operating modes.
iv. It is to be noted that because the engine in hand for case study is port-fuel SI with a three-way catalytic converter, for normal operation mode, lambda (i.e. air-to-fuel ratio over stoichiometric ratio) is only allowed to be within lambda window, namely, between 0.995 and 1. Additionally, since the applied λ-sensor (for the real engine) is ordinary switch-type, lambda must be set to 1. Even if lambda along with other variables were optimized (which was actually tested in this study), the optimal value would be 1. This is why air-fuel ratio is not regarded a variable for part-load mode operation.
v. As for (n_i, θ_i)-based optimization, the rotational dynamic subsystem (Figure 3.1) could be substituted with the respective constant engine speed value, and for each (n_i, θ_i) and different ambient conditions, the values of VVT and spark advance can be optimized.
vi. Exhaust temperature could also be regarded as a constraint.
vii. This requires taking longitudinal dynamic into consideration.
viii. First, anomalous results could be filtered by eliminating the points far-away from the network regression line.
ix. Ensemble averaging leaves the bias error almost unchanged and diminishes variance error.
x. Floating-point operations per second.
xi. Usually, VVT is calibrated to yield maximum volumetric efficiencies, yet other objectives could be adopted.
xii. For this, simple high-level and low-level controllers could also be designed.
xiii. For this procedure, the inverse model (explained in Section 3.3) or the results from preliminary calibration could be utilized.
xiv. The appropriate choice of normalizing factors has a great impact on the quality of convergence.
xv. Other successful methods of map regularization could be found in [2, 10, 11].
xvi. Of course, note that spark advance is adjusted based on the *instantaneous* air mass flow rate into the cylinder (coming from the MAF observer, or the modified signal in the low-level control, to make it more robust and accurate) rather than desired air mass flow rate from the high-level layer (Figure 3.18).

REFERENCES

1. Shamekhi, A.-M. and A.H. Shamekhi, *Engine model-based pre-calibration and optimization for mid-level hierarchical control design*. SAE International Journal of Engines, 2021. **14**(5): p. 651–669.
2. Isermann, R., *Engine Modeling and Control: Modeling and Electronic Management of Internal Combustion Engines*. 2014, Berlin, Germany: Springer.
3. Zhao, J. and M. Xu, *Fuel economy optimization of an Atkinson cycle engine using genetic algorithm*. Applied Energy, 2013. **105**: p. 335–348.
4. Haykin, S., *Neural Networks: A Comprehensive Foundation*. 2nd ed. 1998, Hoboken, NJ, USA: Prentice Hall.
5. Sequenz, H., Emission *modelling and model-based optimisation of the engine control*. In VDI Fortschrittsberichte. Vol. 8. 2013, Düsseldorf, Germany: VDI Verlag.
6. Sequenz, H. and R. Isermann, *Emission model structures for an implementation on engine control units*. IFAC Proceedings Volumes, 2011. **44**(1): p. 11851–11856.
7. Civicioglu, P. and E. Besdok, *A conceptual comparison of the Cuckoo-search, particle swarm optimization, differential evolution and artificial bee colony algorithms*. Artificial Intelligence Review, 2011. **39**: p. 315–346.
8. Kachitvichyanukul, V., *Comparison of three evolutionary algorithms: GA, PSO, and DE*. Industrial Engineering & Management Systems, 2012. **11**(3): p. 215–223.
9. Lim, S.P. and H. Haron, *Performance comparison of genetic algorithm, differential evolution and particle swarm optimization towards benchmark functions*, in *IEEE Conference on Open Systems*. 2013: Sarawak, Malaysia. p. 41–46.
10. Hafner, M. and R. Isermann, *Multiobjective optimization of feedforward control maps in engine management systems towards low consumption and low emissions*. Transactions of The Institute of Measurement and Control, 2003. **25**: p. 57–74.
11. Isermann, R. and H. Sequenz, *Model-based development of combustion-engine control and optimal calibration for driving cycles: General procedure and application*. IFAC-PapersOnLine, 2016. **49**(11): p. 633–640.
12. Heywood, J.B., *Internal Combustion Engine Fundamentals*. 1st ed. 1988, New York, NY, USA: McGraw-Hill.

4 Low-Level Controller Design

Fuel Injection Control

4.1 INTRODUCTION

This chapter concerns the engine fuel injection low-level (feed-back) control. As regards spark advance (feed-back) low-level control, as previously explained, since the engine in hand, just as many other engines, is not equipped with an in-cylinder pressure sensor, only the conventional knock avoidance feed-back controller (described in Section 1.4.3.2) could be utilized, and no further facility can be provided for spark advance feed-back control.

Table 4.1 outlines some of the major problems/challenges (reviewed in Chapter 1) pertaining to fuel injection control, with their corresponding consequences.[i]

Following the arguments in Chapter 1 (Section 1.4.3.1), unlike the practice adopted in some papers, the combination of feed-forward and feed-back controllers (Figure 4.1, re-display of Figure 1.10) is of great significance in achieving both the required bandwidth and robustness for the system.

Furthermore, fixed-gain controllers fail to properly address parameter variations and transient phases, leading to the choice of adaptive control.

Adaptive controllers applied in the literature, as reviewed, are either on-line or off-line. The issue with on-line adaptive controllers is their high computational burden on the ECU, especially at high engine speeds. In other words, as the engine speed rises, the computation time available for ECU (during each cycle) drops, and on-line adaptive controllers may not be feasible at high speeds. Off-line adaptive controller (whose varying parameters are tuned and stored off-line), in spite of faster response, might not be robust, especially as for system ageing and uncertainties pertaining to the feed-forward controller.

In this chapter, after a short glance at the conventional fuel injection control structure, an alternative approach is proposed, which is a Model Predictive Self-tuning Regulator (MP-STR) with adaptive variable functioning (AVF). That is to say, keeping the feed-forward observer in place, self-tuning regulator (STR) is applied to address both parametric and unstructured uncertainties (resulting from errors of the feed-forward controller, parameter variations during transient phases, ageing, etc.), but in a predictive practice, in order to achieve optimal behavior, particularly during transients. Additionally, it is applied with AVF; that is, at lower engine speeds, when the computation time available is longer, the MP-STR is wholly triggered, yet at higher speeds with

DOI: 10.1201/9781003323044-4

TABLE 4.1
Some of the Major Problems/Challenges in the Publications Reviewed with Their Consequences, Regarding Fuel Injection Feed-Back Control

Problems/Challenge	Consequences
Pure feed-back control (without the feed-forward controller/observer)	Bandwidth problem
Fixed gain control	Unfavorable behavior in different transients
Purely on-line adaptive approaches	Not real time at higher speeds
Purely off-line adaptive approaches	Not robust in different operating conditions
Neglecting transient behavior	Adverse emissions during transients

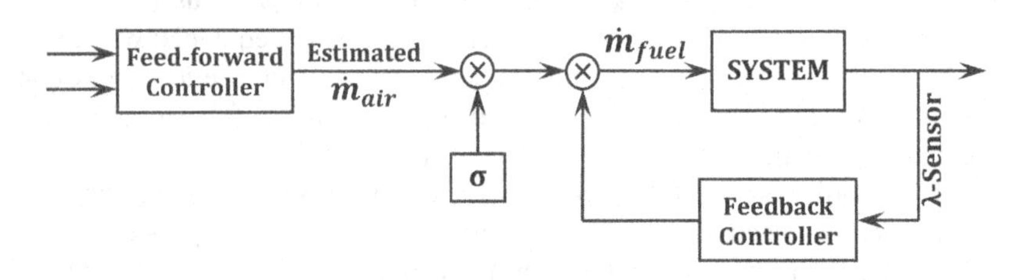

FIGURE 4.1 Re-display of the combination of feed-forward and feed-back controllers.

shorter computation time in hand, a lighter controller (e.g. only the predictive compensator with off-line parameters, possibly coupled with a simple PI controller) is employed. This is in fact the subject of adaptive variable rate (AVR) tasks control [1, 2]. To sum up, as will be seen, the final control structure will be able to appropriately deal with transients and uncertainties (both parametric and unstructured) but in real time. It is only assumed that the system is equipped with a wide-range[ii] (universal) λ-sensor before the three-way (TW) catalytic converter (TWC).

In what follows, after system modeling, the control strategies are explained, and finally, the control system results will be examined [3]. But first, the air mass flow observer is examined. Although it is basically a feed-forward controller, and logically belongs to the mid-level layer, for the sake of coherence and better understanding, it is discussed here.

4.2 AIR MASS FLOW OBSERVER

The feed-forward air mass flow observer could be the combination of intake manifold subsystems of a mean value or discrete event engine model i.e. throttle body, receiver, and gas exchange subsystems. In this structure, the manifold dynamic needs to be considered, and the existing sensor signals (including throttle angle sensor, manifold pressure sensor, etc.) are employed for in-cylinder air mass flow estimation. Figure 4.2 depicts the observer structure, in which X is defined as [4]

$$X = \lambda_l \left(p_m, \omega_e \right) \cdot \frac{V_d}{V_m} \cdot \frac{1}{n_{\text{cyl}}}, \tag{4.1}$$

where $\omega_e(t)$, P_m, V_m, and V_d stand for density of the fluid into the cylinder, engine speed, manifold pressure, manifold volume, and displacement volume, respectively. λ_l is the experimental volumetric coefficient.

Also, p_a and ϑ_a are ambient pressure and temperature, and $\Psi(.)$ is defined as follows:

$$\Psi\left(\frac{P_{\text{in}}(t)}{P_{\text{out}}(t)}\right) = \begin{cases} \sqrt{k\left[\dfrac{2}{k+1}\right]^{\frac{k+1}{k-1}}}, & P_{\text{out}} < P_{\text{cr}} \\ \left[\dfrac{P_{\text{out}}}{P_{\text{in}}}\right]^{1/k} \cdot \sqrt{\dfrac{2k}{k-1} \cdot \left[1 - \left(\dfrac{P_{\text{out}}}{P_{\text{in}}}\right)^{\frac{k-1}{k}}\right]}, & P_{\text{out}} > P_{\text{cr}} \end{cases}. \tag{4.2}$$

In above relations, P_{in} and ϑ_{in}, P_{out}, and k, represent ambient pressure and temperature, intake manifold pressure, and specific heat ratio of air, respectively. P_{cr} as critical pressure is

$$P_{\text{cr}} = \left[\frac{2}{k+1}\right]^{\frac{k}{k-1}} \cdot P_{\text{in}}. \tag{4.3}$$

In Figure 4.2, "A" represents the instantaneous *effective* intake area of throttle body (derived based on throttle angle). C_d, discharge coefficient, is an experimental parameter. τ_{seg} (segment time) is the standard sampling time in ECU.

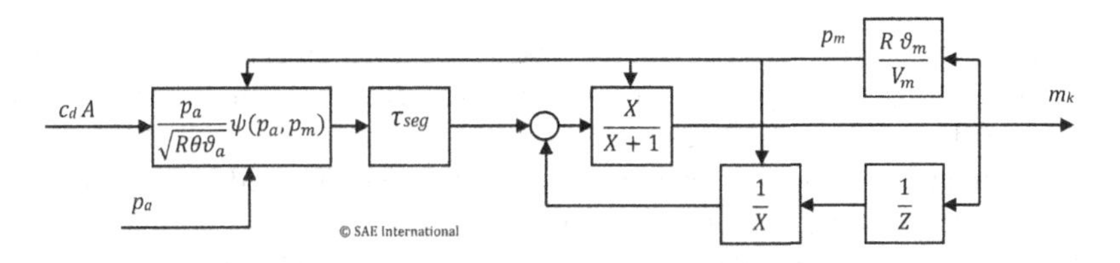

FIGURE 4.2 Block diagram of the feed-forward observer [4].

4.3 MODELING

The major dynamic behavior in the engine fuel path results from the wall-wetting phenomenon. The fuel injected into intake ports does not instantly and totally enter the cylinders but is partly stored in the port walls as puddles of fuel mass.

The fuel from these puddles also evaporates, resulting in the following mass balance equations (Aquino model, Figure 4.3) [4]:

$$\dot{m}_\varphi(t) = (1-\kappa)\cdot \dot{m}_\psi(t) + \frac{m_f(t)}{\tau}$$

$$\frac{d}{dt} m_f(t) = \kappa \cdot \dot{m}_\psi(t) - \frac{m_f(t)}{\tau} \tag{4.4}$$

where $\dot{m}_\psi(t)$ represents the injected fuel mass, $\dot{m}_\varphi(t)$ is the fuel mass flow rate into the cylinder, and $m_f(t)$ is the mass of fuel film. The coefficients κ and τ are functions of engine speed, manifold pressure, fuel temperature, etc., and act as magnifier and time constant for the system. As can be seen, the system entails lag, delay, and parameter variations.

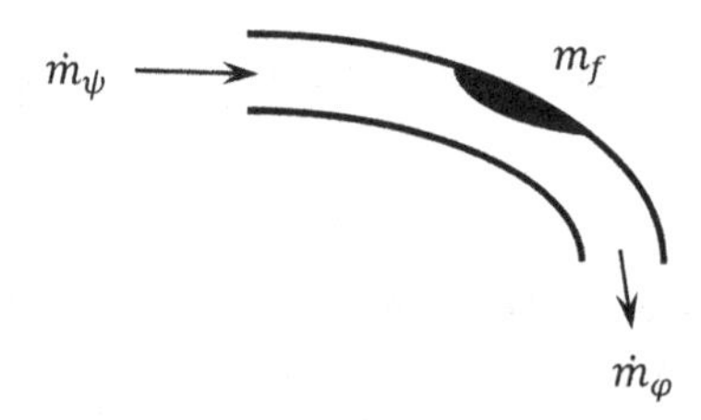

FIGURE 4.3 A simple schematic of wall-wetting.

Defining y = $\dot{m}_\varphi(t)$ and u = $\dot{m}_\psi(t)$:

$$Y(s) = \left((1-\kappa) + \frac{\kappa}{\tau s + 1}\right) U(s). \tag{4.5}$$

In reality, the output "y" is computed based on measurement from a universal (wide-range) λ-sensor signal fed back from downstream of the cylinder. As a result, this feed-back signal is delayed.

Computations in a real engine processing unit are performed in discrete time domain. The sampling time in an ECU (called segment time, τ_{seg}) is the duration between two consecutive plain (i.e. with zero advance or retard) ignition events. As the engine speed dynamic is slow, τ_{seg} could be deemed constant for the relevant period of time[iii] [4].

Equation (4.5) in discrete domain becomes

$$Y(z) = (1-\kappa) + \kappa \cdot \frac{1-e^{-\tau_{seg}/\tau}}{z - e^{-\tau_{seg}/\tau}} \cdot U(z) \tag{4.6}$$

or, in q-domain representation:

$$Y\left(q^{-1}\right)=\frac{(1-\kappa)+\left(\kappa-e^{-\tau_{seg}/\tau}\right)q^{-1}}{\left(1-e^{-\tau_{seg}/\tau}\cdot q^{-1}\right)}\cdot U\left(q^{-1}\right). \tag{4.7}$$

Now, assuming pure *delay d* for λ-sensor,[iv] the output "x" at λ-sensor would be

$$X\left(q^{-1}\right)=\frac{1}{q^{d}}\cdot\frac{(1-\kappa)+\left(\kappa-e^{-\tau_{seg}/\tau}\right)q^{-1}}{\left(1-e^{-\tau_{seg}/\tau}\cdot q^{-1}\right)}\cdot U\left(q^{-1}\right). \tag{4.8}$$

or, in the standard form, it would be:

$$A\left(q^{-1}\right)\cdot Y\left(q^{-1}\right)=q^{-d}\cdot B\left(q^{-1}\right)\cdot U\left(q^{-1}\right), \tag{4.9}$$

where

$$A\left(q^{-1}\right)=1-e^{-\tau_{seg}/\tau}\cdot q^{-1}$$

$$B\left(q^{-1}\right)=(1-\kappa)+\left(\kappa-e^{-\tau_{seg}/\tau}\right)q^{-1}. \tag{4.10}$$

For complete discrete event modeling, sequential injection for multiple cylinders needs to be taken into account. In a port fuel injection engine, the fuel path consists of n_{cyl} (i.e. the number of cylinders) separate wall-wetting dynamics. Interestingly, gas mixing of the corresponding ports in the exhaust manifold evolves in parallel as well. In other words, the inputs to these dynamic elements are imposed in a sequential manner (synchronized to the crank angle), and the output signals are received in the same pattern. Therefore, such a multi-element system can be modeled as a multiplexed system [4]. Accordingly, we have

$$Y(z)=(1-\kappa)+\kappa\cdot\frac{1-e^{-\frac{n_{cyl}\cdot\tau_{seg}}{\tau}}}{z^{n_{cyl}}-e^{-\frac{n_{cyl}\cdot\tau_{seg}}{\tau}}}\cdot U(z) \tag{4.11}$$

where n_{cyl} is the number of cylinders. In q-domain representation, we have

$$X\left(q^{-1}\right)=\frac{1}{q^{d}}\cdot\frac{(1-\kappa)+\left(\kappa-e^{-\frac{n_{cyl}\cdot\tau_{seg}}{\tau}}\right)q^{-n_{cyl}}}{\left(1-e^{-\frac{n_{cyl}\cdot\tau_{seg}}{\tau}}\cdot q^{-n_{cyl}}\right)}\cdot U\left(q^{-1}\right). \tag{4.12}$$

For the standard form in (4.9), we have

$$A\left(q^{-1}\right)=1-e^{-\frac{n_{\text{cyl}}\cdot\tau_{\text{seg}}}{\tau}}\cdot q^{-n_{\text{cyl}}}$$

$$B\left(q^{-1}\right)=(1-\kappa)+\left(\kappa-e^{-\frac{n_{\text{cyl}}\cdot\tau_{\text{seg}}}{\tau}}\right)q^{-n_{\text{cyl}}}. \tag{4.13}$$

It should be noted that the system is the "minimum phase."[v]

4.4 CONTROL

4.4.1 Conventional Feed-Back Control

Ordinarily, one switch-type lambda sensor is placed before catalytic converter, whose voltage drops from less than about 1000 mV for rich mixture, down to about 0 mV for lean mixture, which obviously resembles a binary signal. At stoichiometry, the voltage is 450 mV. The sensor signal is fed back through a binary filter and then a PI controller (Figure 4.4).

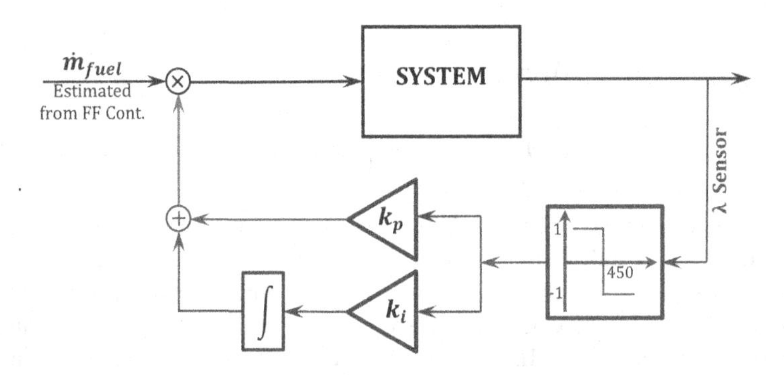

FIGURE 4.4 Block diagram of fuel injection feed-back controller [4].

Note that the feed-back controller signal is multiplied by the feed-forward fuel injection signal. This is due to multiplicative nature of errors in the system. Moreover, because of asymmetric behavior of the switch-type λ-sensor, the control system shows a bias in the mean value of air-fuel ratio towards lean conditions. To compensate, the switch from rich to lean in the binary filter is delayed by t_v [4].

The parameters k_i, k_p, and t_v ought to be tuned for every operating point and stored in look-up tables in the ECU to be recalled during on-line control. Nonetheless, this off-line adaptive strategy is obviously not robust.

What is more, the above structure clearly does not take the dynamic behavior of wall-wetting subsystem into consideration (as can be seen in the response in Figure 1.10). In some design structures, the inverse of wall-wetting dynamic is incorporated. Apart from realization considerations, this compensation is also not robust.

4.4.2 Alternative Feed-Back Control

Feed-forward observer provides an estimated baseline signal for actuators (injectors). As will be further examined in Section 4.5, the error of this estimation could be deemed as multiplicative unstructured uncertainty. Furthermore, due to delay, lag, and parameter variations of the wall-wetting subsystem, the aforesaid baseline signal needs to be adaptively compensated to yield appropriate $\dot{m}_\varphi$ at the output of the system.

The goal of this section is to exploit MP-STR as feed-back controller, so as to address unstructured and parametric uncertainties and provide optimal control behavior. Figure 4.5 displays a simple schematic of MP-STR. As can be seen, it is actually an STR, but with model predictive control as the controller block. At each time step, first, an on-line system identification is performed by Recursive Least Square (RLS),[vi] and afterward, the identified system parameters ($\hat{A}$ and $\hat{B}$, as estimations of A and B, defined in Section 4.3) are used for solving the Diophantine equation. The solution of the Diophantine equation is then utilized for tuning the parameters of the model predictive controller block. As will be explained, for lower computational burden, a *direct* STR (without having to solve the Diophantine equation) is employed.

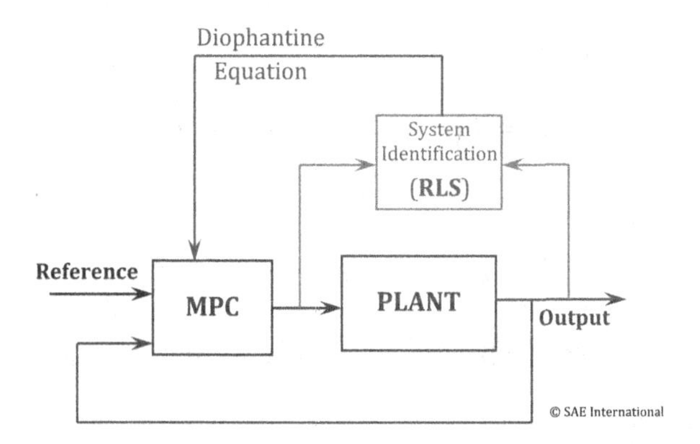

FIGURE 4.5 A simple schematic of MP-STR [3].

As mentioned, first an RLS on-line system identification is conducted from previous inputs and outputs to estimate $\hat{A}$ and $\hat{B}$. Now let [5–7]:

$$A\left(q^{-1}\right) = 1 + A_1 \cdot q^{-1} + \cdots + A_n \cdot q^{-n}$$

$$B\left(q^{-1}\right) = q^{-d} \cdot \left(B_1 \cdot q^{-1} + \cdots + B_m \cdot q^{-m}\right). \tag{4.14}$$

Using (4.9), we obtain

$$y(t)=\left[-y(t-d)\ldots-y(t-n-d+1)\ u(t-d)\ldots u(t-m-d+1)\right]\begin{bmatrix} A_1 \\ \vdots \\ A_n \\ B_1 \\ \vdots \\ B_m \end{bmatrix}. \tag{4.15}$$

Namely

$$y(t)=\Phi^T(t-1)\cdot\Theta \tag{4.16}$$

Then

$$\hat{\Theta}(t)=\hat{\Theta}(t-1)+K(t)\cdot\left[y(t)-\Phi^T(t)\cdot\hat{\Theta}(t-1)\right], \tag{4.17}$$

where

$$K(t)=P(t-1)\cdot\Phi(t)\cdot\left[I+\Phi^T(t)\cdot P(t-1)\cdot\Phi(t)\right]^{-1}$$

$$P(t)=\left[I-K(t)\cdot\Phi^T(t)\right]\cdot P(t-1). \tag{4.18}$$

Now, for the Diophantine equation:

$$1=F\left(q^{-1}\right)\cdot\hat{A}\left(q^{-1}\right)+q^{-d}\cdot G\left(q^{-1}\right), \tag{4.19}$$

where

$$G\left(q^{-1}\right)=\alpha_0+\alpha_1\cdot q^{-1}+\cdots+\alpha_{n-1}\cdot q^{-(n-1)}=\alpha\left(q^{-1}\right)$$

$$F\left(q^{-1}\right)\cdot\hat{B}\left(q^{-1}\right)=\beta_0+\beta_1\cdot q^{-1}+\cdots+\beta_{m+d-1}\cdot q^{-(m+d-1)}=\beta\left(q^{-1}\right), \tag{4.20}$$

the following equation holds:

$$y(t+d)=\alpha\left(q^{-1}\right)\cdot y(t)+\beta\left(q^{-1}\right)\cdot u(t), \tag{4.21}$$

which is similar to the original system equation (4.9), and thus, could be estimated directly as (4.15) and (4.16), for direct STR.[vii]

Now, since the magnitude of the system inputs is not of concern, to minimize the optimal cost function:

$$J(t+d)=\left\{\tfrac{1}{2}\left(y(t+d)-y_d(t+d)\right)^2\right\}, \tag{4.22}$$

the optimal input would be

$$u(t) = \frac{\left\{y_d(t+d) - \alpha(q^{-1})\cdot y(t) - \beta'(q^{-1})\cdot u(t-1)\right\}}{\beta_0}, \tag{4.23}$$

where

$$\beta'(q^{-1}) = q\cdot\left[\beta(q^{-1}) - \beta_0\right]. \tag{4.24}$$

For each time step τ_{seg}, a new $u(t)$ is computed and imposed to the system.

4.5 FURTHER DISCUSSION REGARDING THE SYSTEM UNSTRUCTURED UNCERTAINTIES

In this section, multiplicative unstructured uncertainty is to be investigated. It will be demonstrated that for a multiplicative uncertainty of ε, if the calculated output $\boldsymbol{y}$ (calculated based on the measured output λ and the estimated value of $\widehat{\dot{\boldsymbol{m}}_a}$) is regulated to the estimated reference $\hat{\boldsymbol{y}}_d$ (by means of the MP-STR controller), then the real output $\boldsymbol{y}'$ will be regulated to the real desired $\boldsymbol{y}_d$ (and thus, the measured output λ will be regulated to its desired value 1). Simply put, although the system output is regulated based on an estimation, it is valid (Figure 4.6).

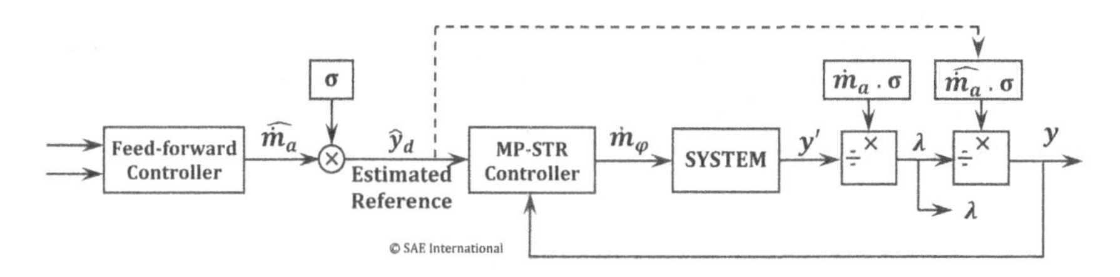

FIGURE 4.6 Fuel injection control structure with feed-forward and feed-back loops [3].

Here, let

$\dot{m}_a$: Actual air mass flow rate into the cylinder
$\widehat{\dot{m}_a}$: Estimated air mass flow rate into the cylinder
y': Actual fuel mass flow rate into the cylinder
y: Estimated air mass flow rate into the cylinder (i.e. estimated output)
$\hat{y}_d$: Estimated desired (reference) value for y
y_d: Actual desired (reference) value for y'
λ: Lambda, the ratio of actual air-fuel ratio (AFR) to the stoichiometric AFR, measured output
φ: Equivalence ratio, defined as $1/\lambda$.

Also

$$y_d = \dot{m}_a \cdot \sigma,$$

$$\hat{y}_d - \widehat{\dot{m}_a} \cdot \sigma, \tag{4.25}$$

$$\varepsilon = \widehat{\dot{m}_a} \big/ \dot{m}_a \equiv \text{Multiplicative unstructured uncertainty.} \tag{4.26}$$

Then

$$y_d = \hat{y}_d \big/ \varepsilon\,. \tag{4.27}$$

As mentioned, we would like y' to converge to y_d. Namely

$$y' \rightarrow y_d. \tag{4.28}$$

The controller actually makes the calculated output y to converge to the estimated reference value of $\hat{y}_d$. Namely

$$y \rightarrow \hat{y}_d. \tag{4.29}$$

Now, for the measured output $\boldsymbol{\lambda}$, the calculated output $\boldsymbol{y}$ (which is fed back to the controller) is estimated by

$$y = \frac{\widehat{\dot{m}_a}}{\lambda} \cdot \sigma. \tag{4.30}$$

Also, for the actual values

$$1 \big/ \lambda = \frac{y'}{\dot{m}_a \cdot \sigma}. \tag{4.31}$$

Therefore, substituting (4.31) into (4.30), we obtain

$$y = y' \frac{\widehat{\dot{m}_a}}{\dot{m}_a}. \tag{4.32}$$

Now, from (4.26) and (4.32), we obtain

$$y = y'\ \varepsilon \tag{4.33}$$

or

$$y' = y \big/ \varepsilon\,. \tag{4.34}$$

Now, if the controller makes y to converge to $\hat{y}_d$, that is, if (4.29) holds, then from (4.34), we obtain

$$y' = \hat{y}_d / \varepsilon. \tag{4.35}$$

Then, from (4.27) and (4.35), we obtain

$$y' = y_d \tag{4.36}$$

which is the very goal demanded in (4.28).

Therefore, for the multiplicative uncertainty of ε, if the controller regulates $y \to \hat{y}_d$, then $y' \to y_d$.

Moreover, from (4.33), for the multiplicative unstructured uncertainty of ε, the system equation becomes

$$A(q^{-1}) \cdot Y(q^{-1}) = \varepsilon \cdot q^{-d} \cdot B(q^{-1}) \cdot U(q^{-1}). \tag{4.37}$$

4.6 RESULTS

In this section, the control system results are presented. To better demonstrate the performance of the designed controller (MP-STR), first parametric uncertainties are imposed and afterwards the capabilities of the control system in tackling both parametric and unstructured uncertainties will be put to the test. In other words, it is first presumed that the baseline signal provided by feed-forward controller is accurate and merely parametric uncertainties exist. Afterward, the general case will be examined, and the control system will tackle baseline signal errors, along with lag, delay, and parameter variations of the system.

In all the results to come, the control output $\dot{m}_\varphi(t)$ (or $M_\varphi(t)$, namely, the real fuel flow into the cylinder) is to be adjusted by means of the control input $\dot{m}_\psi(t)$ (or $M_\psi(t)$, namely, the injected fuel into the intake runner) such that the real λ (after the time delay d) is regulated to the stoichiometric value 1, in the presence of uncertainties.

Note that excessive uncertainties will be imposed so as to challenge the control system capabilities. They might seem not realistic, yet they are to take worst cases into consideration.

4.6.1 Parametric Uncertainties

Figure 4.7 portrays the subsystem's parameter variations over different operating conditions. The controller is, of course, not aware of these parameter variations. The control system results are displayed in Figure 4.8. The reference and

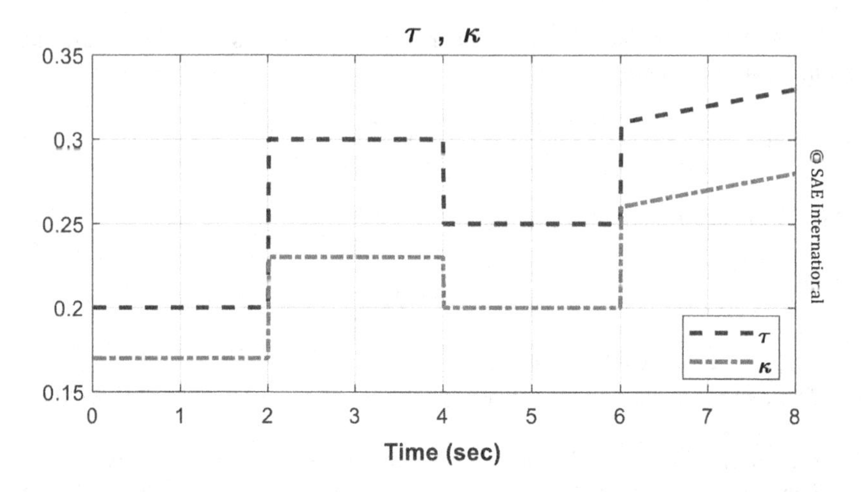

FIGURE 4.7 The subsystem's parameter variations over different operating conditions [3]. Although excessive, the inputs are to demonstrate the control system capabilities. The intervals of parameters are derived from [4].

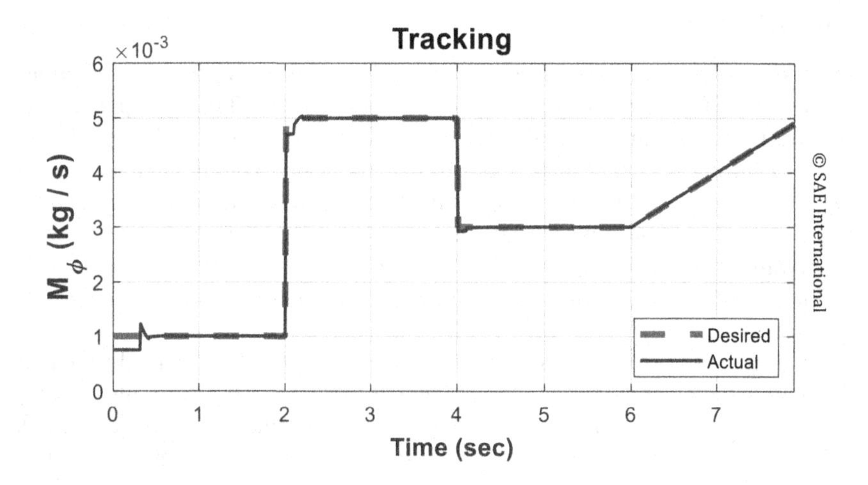

FIGURE 4.8 The control system response [3].

the parameter variations might seem radical (particularly at 6 s), but they are to challenge the control system and demonstrate its capabilities. As can be seen, the desired output $\dot{m}_{\varphi}(t)$ is properly tracked, even for sharp transient reference and parameter variations. The system identification starts with incorrect initial value but readily converges to correct values in a matter of 0.4 s. Figure 4.9 indicates the resulting lambda (λ) of the mixture entering the cylinder. Figure 4.10 depicts the injected fuel commanded by the controller.

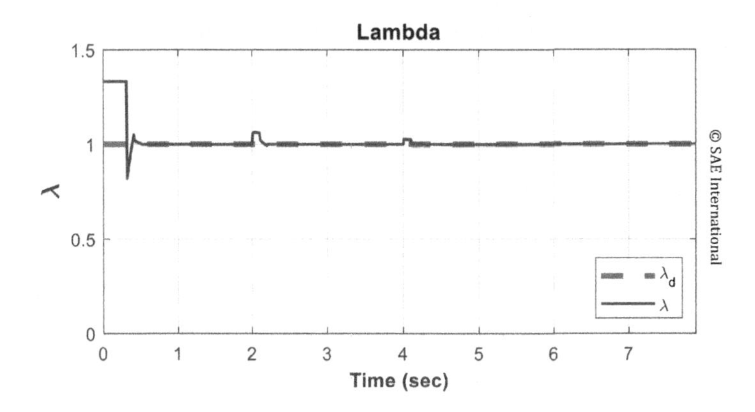

FIGURE 4.9 Lambda variations over the imposed transient conditions [3].

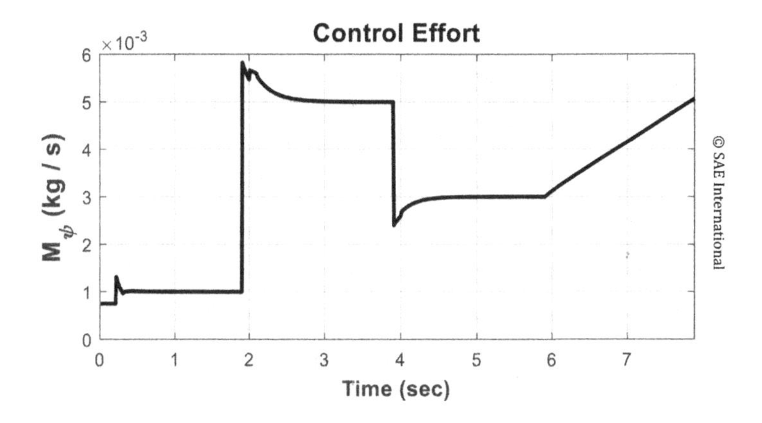

FIGURE 4.10 The control command (i.e. the injected fuel) [3].

4.6.2 Parametric and Unstructured Uncertainties

As previously mentioned, the control system considered here is a feed-back controller modifying the baseline signal of a feed-forward controller. The feed-forward controller is actually an air mass flow observer/estimator. In the previous section, it was presumed that there only exist parametric uncertainties and the feed-forward observer provides exact set points for the feed-back controller.

In this section, both parametric and unstructured uncertainties are regarded. The system unstructured uncertainties could be deemed as multiplicative uncertainties lumped into the observer output [4].

Accordingly, assuming an unstructured uncertainty of "ε %," as indicated in Figure 4.11,[viii] and parametric variations as depicted in Figure 4.7, the control system performance would be as represented in Figure 4.12. Figure 4.13 displays the

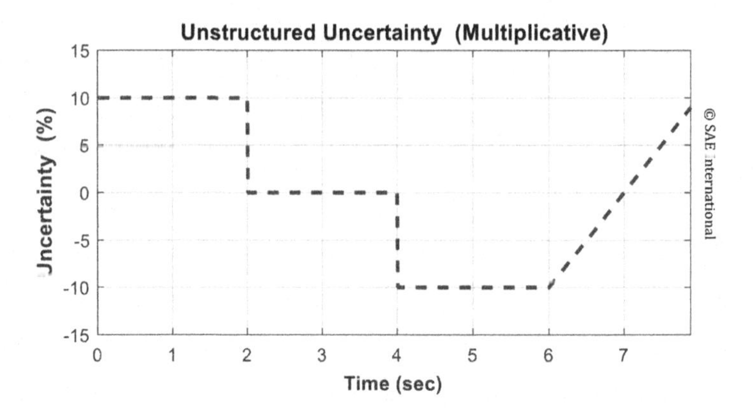

FIGURE 4.11 The unstructured multiplicative uncertainty [3].

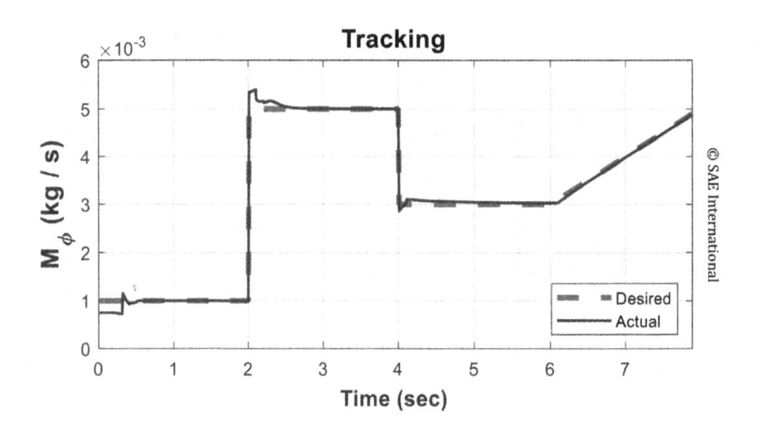

FIGURE 4.12 The control system response [3].

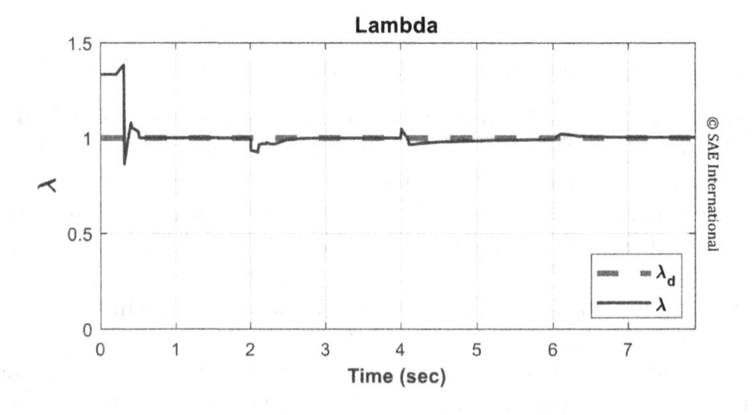

FIGURE 4.13 Lambda variations over the imposed transient conditions [3].

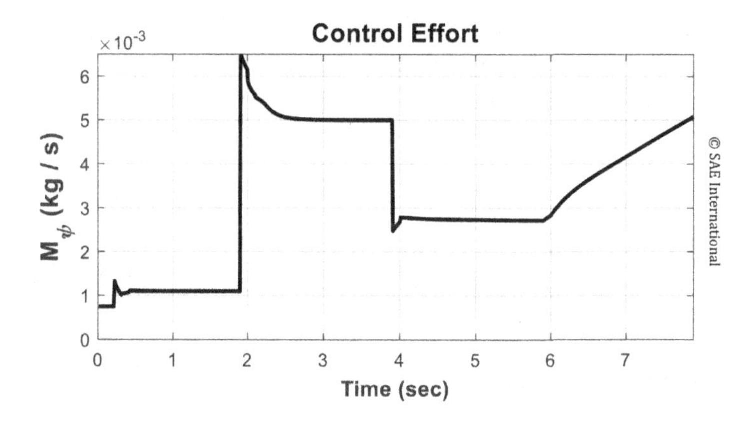

FIGURE 4.14 The control command (i.e. the injected fuel) [3].

resulting lambda (λ) of the mixture entering the cylinder. Figure 4.14 represents the injected fuel commanded by the controller.

To further demonstrate the appropriate performance of the control system, the performance of an ordinary PI fuel injection controller, for the same uncertainties is indicated in Figure 4.15. As can be seen, the MP-STR control system shows superior capability in dealing with the unstructured and parametric uncertainties imposed.

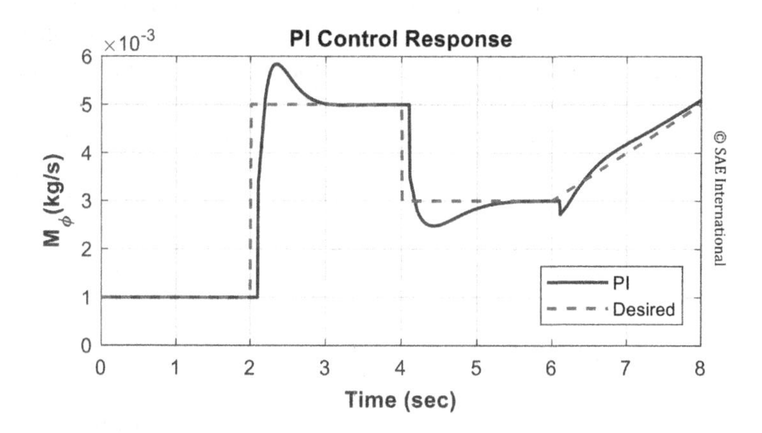

FIGURE 4.15 The performance of an ordinary PI fuel injection controller, for comparison [3].

4.6.3 Adaptive Variable Functioning

At high engine speeds, there might be not enough time for the ECU to cope with the computational burden of MP-STR. In these circumstances,[ix] the control system can be switched to an offline-adaptive predictive controller. That is, the RLS is unloaded, and the predictive controller is run based on the pre-stored system parameters. To address the control system uncertainties and system ageing, the controller is coupled with a simple PI controller (MP-PI). Figure 4.16 portrays the structure of MP-PI.

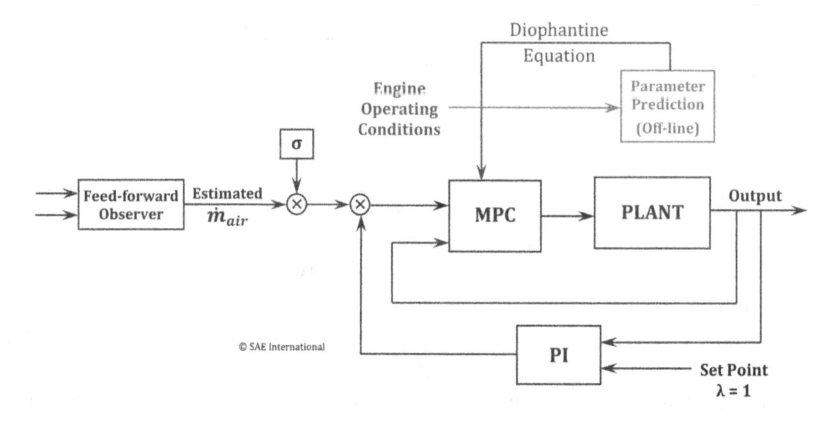

FIGURE 4.16 The structure of MP-PI controller [3].

The control system response is displayed in Figure 4.17. As can be observed, in phases II and IV, the controller switches from MP-STR to MP-Offline-STR + PI. The system parametric and unstructured uncertainties are just as before, but with 20% of error in the system parameters stored for the off-line MP controller. Figure 4.18 represents the respective control command. As can be seen, the performance is not as good as that of pure MP-STR, but still better than Figure 4.15.

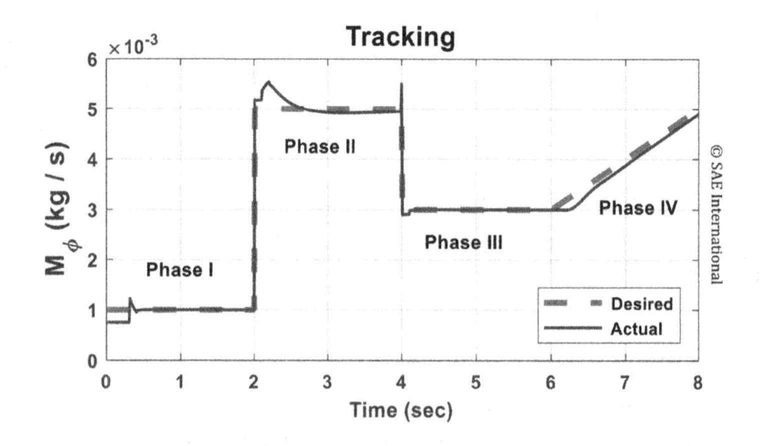

FIGURE 4.17 The system response for AVF controller [3].

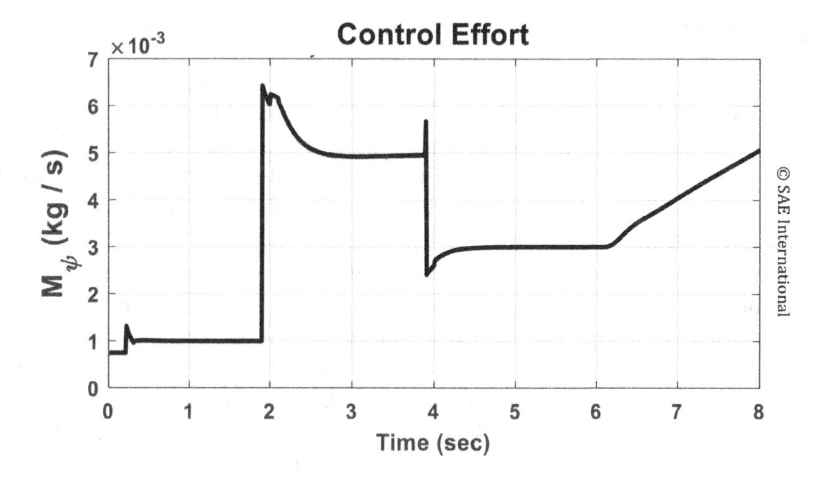

FIGURE 4.18 The AVF control command [3].

4.7 OUTER LOOP CONTROL

So far, we attempted to enhance AFR control in transients. Nonetheless, in order to fully make sure that lambda is kept within λ-window during transients, an outer control loop with an after catalyst λ-sensor is required. There are special catalytically active materials (e.g. cerium) applied in TWC, so as to incorporate an oxygen reservoir in it. The duty of this reservoir is to cope with temporary excursions to lean or rich lambda sides during transients. To realize this, the oxygen storage level must be kept at half the available capacity [4]. Since the oxygen storage level cannot be measured directly, an extra λ-sensor is installed downstream of TWC, in order to build up an observer.

To model the storage behavior of TWC, an integrator is incorporated into the system model. The upstream λ-sensor is of wide-range type, yet the downstream one is a switch-type λ-sensor. Figure 4.19 depicts the schematic of the inner and outer loop controllers.

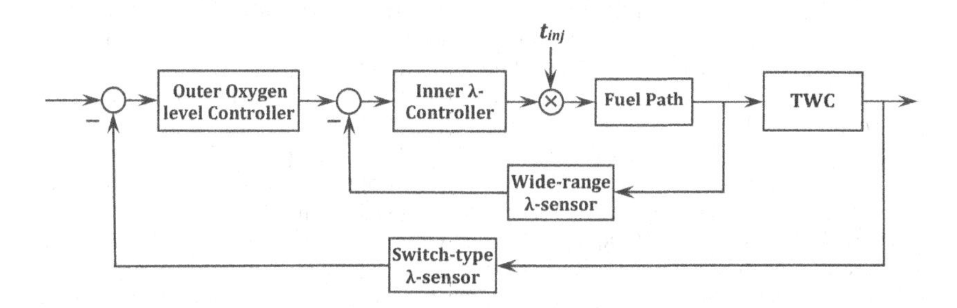

FIGURE 4.19 Schematic of the inner and outer loop controllers [4].

4.8 SUMMARY

In this chapter, engine fuel injection control was examined, and after reviewing the feed-forward observer and conventional feed-back controllers, an MP-STR with AVF was proposed for fuel injection feed-back control. As mentioned, the major problems in fuel injection control are transients and uncertainties. STR was applied to address both parametric and unstructured uncertainties (caused by parameter variations in transient phases, errors of the feed-forward controller, ageing, etc.), but in a predictive practice, so as to attain optimal behavior, especially in transient phases. Additionally, it was accompanied with AVF, in order to maintain the control system work in real time; that is, at lower engine speeds with longer computation time available for ECU, the MP-STR was wholly triggered, yet at higher engine speeds with shorter computation time in hand, RLS was unloaded and an off-line adaptive model predictive controller coupled with a simple PI controller (i.e. off-line adaptive MP-PI) was applied.

As observed in the results, although overly challenging unstructured and parametric uncertainties were applied, MP-STR was well capable of dealing with transients, much more preferable than the prevalent PI controllers. With regard to the off-line-adaptive MP-PI, although not as good as MP-STR, the results were still better than that of the pure PI controller.

In the end, the outer loop control for oxygen-level regulation in TWC was introduced. As mentioned, the feed-back controller presented above would work as an inner loop for AFR control, yet another controller with an after catalyst (switch-type) λ-sensor would form an outer loop, in order to regulate the oxygen-level in TWC at half the capacity, and thus, to fully ensure transients staying within λ-window.

NOTES

i. Not all the papers reviewed suffer from all the problems mentioned.
ii. Unlike switch-type λ-sensor, which only detects lean or rich mixtures, this type is capable of indicating the exact value of lambda.
iii. In fact, the rotational dynamic (directly affecting the engine speed) is at least 10 times slower than the intake manifold dynamic. Hence, the engine speed could be considered constant for the relevant period of time. Therefore, the control problem could be solved in discrete time domain, instead of discrete event domain [1].
iv. Note that λ-sensor is usually modeled as a pure delay. However, a more advanced model (with lag) could be added although it makes the system modeling more complicated.
v. A system with right-hand-plane zero is called "minimum phase." For SISO systems, it causes undershoot, yet, for MIMO systems it manifests a combination of undershoot and/or high interactions, which are all control difficulties.
vi. RLS is a powerful technique in on-line system identification based on recursively estimating the coefficients of a linear weighted function such that sum of the squared errors of identification are minimized. Just as least square, it is unbiased. It is in fact a Kalman filter.

vii. In other words, $\alpha(q^{-1})$ and $\beta(q^{-1})$ could be directly estimated from the system's previous inputs and outputs, using (4.17) and (4.18), but with α_i and β_i instead of A_i and B_i.
viii. Maximum of error resulting from the feed-forward observer in [8] is around 5%. Here, to challenge the control system, and thus, to demonstrate its capabilities, a higher maximum error, i.e. 10%, is considered.
ix. The exact conditions (e.g. engine speed, etc.) where switching should happen depends on the hardware specifications of the ECU and the FLOPS of other functions running and must be specifically determined for the respective engine control system. The principal source of computational burden belongs to RLS (which is unloaded at higher speeds) and could have multiplicative FLOPS of 588. Other parts involve much smaller FLOPS, for e.g., 16 additives and 18 multiplicatives. Note that the computational time available for ECU can be up to 10 times longer at lower speeds (compared to higher speeds).

REFERENCES

1. Biondi, A. and G. Buttazzo, *Engine control: Task modeling and analysis*, in *2015 Design, Automation & Test in Europe Conference & Exhibition*. 2015. p. 525–530.
2. Biondi, A. and G. Buttazzo, *Modeling and analysis of engine control tasks under dynamic priority scheduling*. IEEE Transactions on Industrial Informatics, 2018. **14**(10): p. 4407–4416.
3. Shamekhi, A.-M. and A.H. Shamekhi, *Engine adaptive fuel injection control using model predictive self-tuning regulator with adaptive variable functioning*. SAE International Journal of Engines, 2021. **14**(5): p. 671–682.
4. Guzzella, L. and C. Onder, *Introduction to Modeling and Control of Internal Combustion Engine Systems*. 2nd ed. 2010, Berlin, Germany: Springer.
5. Åström, K.J. and B. Wittenmark, *Adaptive Control*. 2nd ed. 1995, Mineola, NY, USA: Dover Publications.
6. Camacho, E.F. and C. Bordons, *Model Predictive Control*. 2007, London, UK: Springer.
7. Goodwin, G.C. and K.S. Sin, *Adaptive Filtering Prediction and Control*. 2009, Mineola, NY, USA: Dover Publications, Inc.
8. Stotsky, A. and I. Kolmanovsky, *Application of input estimation techniques to charge estimation and control in automotive engines*. Control Engineering Practice, 2002. **10**: p. 1371–1383.

5 High-Level Controller Design

Torque Control

5.1 INTRODUCTION

In this chapter, the engine high-level torque controller is designed. As mentioned in Chapter 1, in conventional electronic control units (ECUs), the demanded torque is put through a first-order filter (before submitting to the mid-level layer) to avoid a sudden rise or fall, and to yield a gentle variation. Although of proper bandwidth and fast response, this feed-forward torque control approach is not robust (and probably, not accurate). It is also not optimal during transients.

Table 5.1 outlines some of the major problems/challenges in the literature reviewed (in Chapter 1) with their corresponding consequences, regarding engine torque control.

The articles reviewed typically ignore the hierarchical control structure and merely focus on feed-back controllers. It shall be noted that an engine is a complicated multi-input multi-output system and ignoring the hierarchical structure results in high interactions. Apart from this, totally ignoring feed-forward controllers and merely relying on feed-back control is highly unlikely to provide real-time responses and the bandwidth required. It needs to be noted that the system is delayed and also the computation time available for ECU in each engine cycle is extremely short.[i] Additionally, feed-back control is of lower bandwidth compared to feed-forward control. This problem escalates as the feed-back controller becomes more complicated, further degrading the agility of the control system. Consequently, mere feed-back controllers (especially complicated ones) are not practical, and the hierarchical structure, with the combination of feed-forward and feed-back controllers needs to be applied.

On top of these, feed-back controllers (which are of lower bandwidth) basically concern steady-state responses (i.e. low-frequency response) and may fail to account for transients (which are high-frequency responses). As mentioned, a large proportion of standard driving cycles (especially the newer ones) occur in transient states. This escalates when it comes to hybrid electric vehicles. As a matter of fact, in hybrid electric vehicles, the engine has to be frequently started off and loaded up to desired torques, raising the importance of transient operation control.[ii] In other words, the transition path to the desired torque significantly affects the overall amount of fuel consumed[iii] (and even the overall pollutant emission produced).

DOI: 10.1201/9781003323044-5

TABLE 5.1
Some of the Major Problems/Challenges in the Publications Reviewed with Their Consequences, Regarding Engine Torque Control

Problem/Challenge	Consequences
Pure feed-back control	Bandwidth problem
Neglecting hierarchical structure	High interactions
Neglecting optimal transients	Not optimal fuel consumption

In this chapter, an alternative approach for torque control will be presented, which is capable of transient optimal control. In light of the aforementioned arguments, the hierarchical structure is maintained, and to account for transients, a semi-feed-forward torque controller is incorporated in the high-level layer, which, besides bandwidth considerations, provides the optimal transition path to the desired torque. Note that this torque controller in fact replaces the classical first-order filter, after torque demand and torque coordination segments (Section 1.4.1, and Figure 1.8).

In what follows, after a short glance at our proposed methodology, the system modeling and control will be discussed [1]. We believe that this idea can notably improve engine control system performance and would gratefully welcome suggestions by scholars and engineers to improve this approach.

5.2 METHODOLOGY

As argued in the previous section, transient torque control, particularly in hybrid electric vehicles, is a matter of great significance and a semi-feed-forward controller is to be incorporated in the high-level control layer to provide the bandwidth invoked. As illustrated in Figure 5.1, due to the special forms of brake-specific fuel consumption (BSFC) curves in a spark ignition (SI) engine, transition from the starting (current steady-state) point A to the desired (steady-state) point B with the demanded torque could be realized with different paths, each of which entails different overall (cumulative) fuel consumptions.

As a matter of paramount importance, for each starting point[iv] (T_0, n_0) with respective operating conditions, the desired final torque T_1 is inevitably realized with a final engine speed n_1. In other words, besides engine torque T_1, the engine speed n_1 for the destination point B is also definite and could not be assigned arbitrarily. This comes from the fact that at any steady-state operating point (e.g. point A or B), the produced torque (e.g. T_0 or T_1) has to be balanced with the respective external loads. External load emanates from different sources such as road resistance, road slope, aerodynamic drag (or air resistance), etc.[v] Without losing generality, it could be assumed that it is only aerodynamic drag (which basically depends on vehicle speed) that changes during transition, and other load sources remain constant (during transition) from steady-state points A to B. Moreover, for the time being, it is assumed that there is no gear shifting during

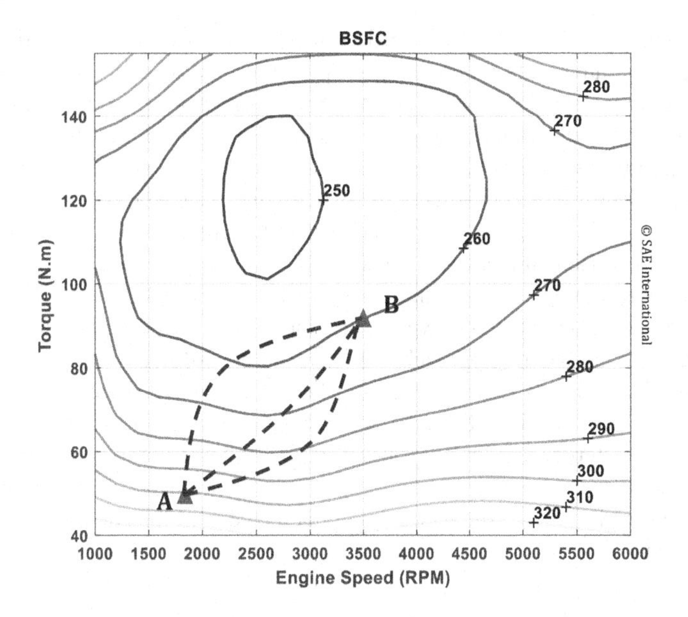

FIGURE 5.1 Different transition paths from point A to point B over a BSFC map [1].

transition. Consequently, the external load from A to B changes *parabolically* (as the aerodynamic drag force is a polynomial of degree 2 with respect to engine speed, in a chosen gear). Thus, the final engine speed n_1 is determined as the point with load torque T_1 in the relevant *parabolic gear curve* (crossing) from point A (Figure 5.2), completely specifying the destination point B.

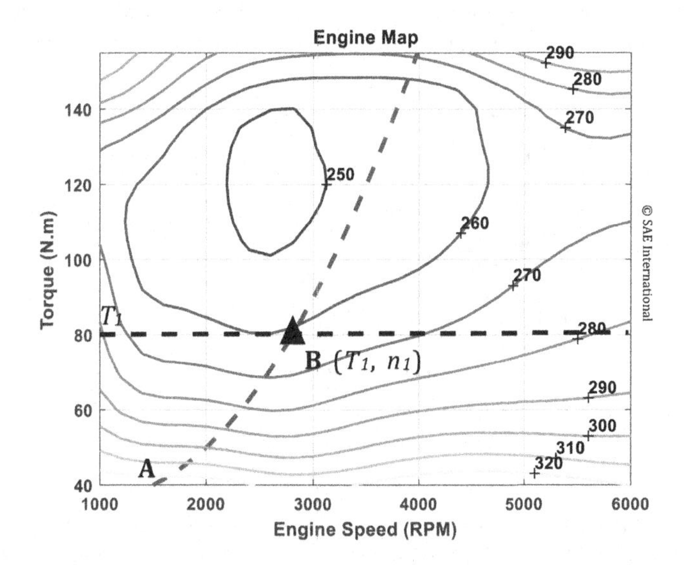

FIGURE 5.2 Intersection of the demanded torque T_1 with aerodynamic transient load curve for the fifth gear from point A [1].

Interestingly enough, hybrid electric vehicles tend to deal with a limited number of steady-state points. Consequently, optimal transition paths could be derived off-line and stored in the high-level layer[vi] without memory restrictions. In this way, not only would transitions with optimal fuel consumption be realized but also bandwidth and memory considerations could be satisfied. For a large number of operating points (as is the case in ordinary vehicles), the optimal curves could be turned into polynomials, and their coefficients could be stored in look-up tables or other efficient black-box identifiers.

In order to conduct the optimization, provide bandwidth, and take the system nonlinearity into account, explicit nonlinear model predictive control will be employed. In this control approach, off-line solutions of nonlinear model predictive control are derived and stored, and hence, a semi-feed-forward controller with large bandwidth is achieved. Yet, owing to memory restrictions of ECU, instead of all-step receding horizon, (what we call) one-step receding horizon will be adopted.

Here, in what follows, the methodology is introduced for one specific case, as an example.

5.3 MODELING

One of the advantages of hierarchical structure is that it involves only one control command from the high-level to mid-level layer. In other words, aside from the engine speed (fed back from an rpm sensor) and ambient conditions, it is based on the air mass flow rate demand (or torque demand) from high-level layer that all the actuator control inputs are determined in the mid-level feed-forward controllers. In this way, the complexity of control system design substantially declines.

As a result, a virtual system consisting of mid-level controllers, low-level controllers (if possible), and of course, the engine could be regarded (Figure 5.3).

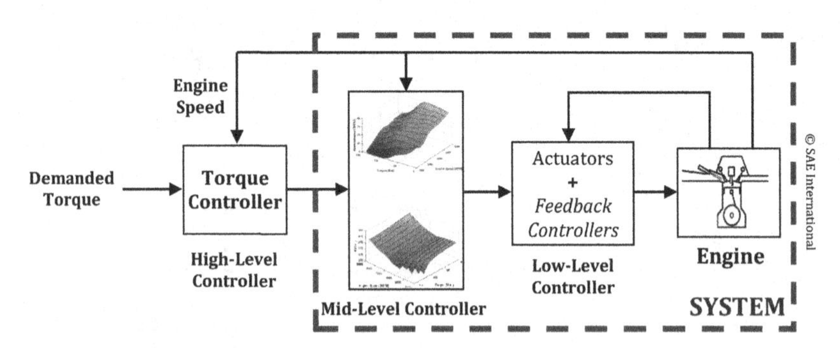

FIGURE 5.3 Schematic of the system and the torque controller [1].

With adequate accuracy, the engine could be deemed as a system of order two, with two dynamics of the intake manifold and rotational inertia. As a matter of fact, the dynamic of the intake manifold is at least 10 times faster than that of the rotational inertia [2]. It is to be noted that torque generation delay (τ_{IPS}, Equation 2.13, Section 2.4.5) is also of the order of crank-shaft revolution, and hence, much longer than the intake manifold's time constant. Accordingly, for the sake of simplicity and reducing the computation time, the intake manifold dynamic could be ignored with decent accuracy, and the engine would be deemed as a system of order one. What is more, supposing proper performance of low-level controllers, they could be ignored just as well. In other words, if low-level feed-back controllers work properly, it could be assumed that there are no uncertainties, and the mid-level controllers are enough for the system in Figure 5.3. Needless to say, should powerful computer hardware be available, the above-ignored subsystems could all be taken into consideration.

Now, separating the rotational inertia subsystem from the engine, and merging the rest with the mid-level control layer, the system could be simplified to a Hammerstein model (Figure 5.4). That is to say, the system model can be simplified to one nonlinear static model feeding a linear dynamic model. This simplification immensely accelerates the computations.

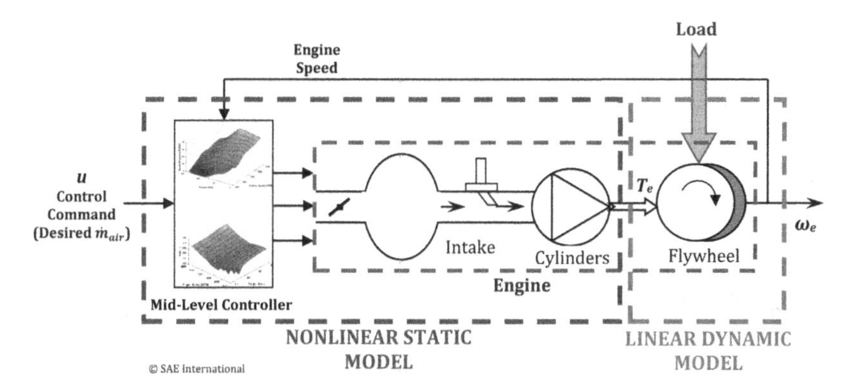

FIGURE 5.4 The resulting simplified Hammerstein model of the system [1].

For the Hammerstein model achieved, the inputs to the nonlinear static model are the desired air mass flow rate into the cylinders (u), and the engine speed (n) and the output would be the produced torque[vii] (T). The inputs of the linear dynamic model consist of the produced torque and the external load (L), and the output would be the resulting engine speed. That is, for the $(k+1)$ th step, we have

$$
\begin{aligned}
T_{k+1} &= f(u_k, n_k) \\
n_{k+1} &= n_k + \delta t \cdot \frac{30}{\pi} \cdot \frac{1}{M_{eq}} \cdot [T_{k-d} - L_k]
\end{aligned}
\tag{5.1}
$$

where $f(\cdot)$ is the nonlinear function representing the nonlinear static subsystem, M_{eq} is the equivalent vehicle mass exerted to the crankshaft,[viii] δt is the time increment, and d denotes the induction-to-torque generation delay.

Regarding external load, as previously explained, it could be appropriately assumed that the load varies parabolically (with respect to engine speed) for a determined gear (Figure 5.2). More details pertaining to force-load modeling is provided in Section 5.4.

The problem as to load modeling, however, is the fact that it entails uncertainty. Yet, since the starting point (i.e. the current state, for e.g. point A) is definite, the uncertainty could be somehow assumed multiplicative. The actual external load curve would be within the band (i.e. uncertainty template) depicted in Figure 5.5. The target point B would be somewhere within the line B′B″. Note that the uncertainty template in Figure 5.5 is exaggerated for a better appearance. A separate comprehensive study is required regarding uncertainty templates, particularly in a hybrid vehicle.

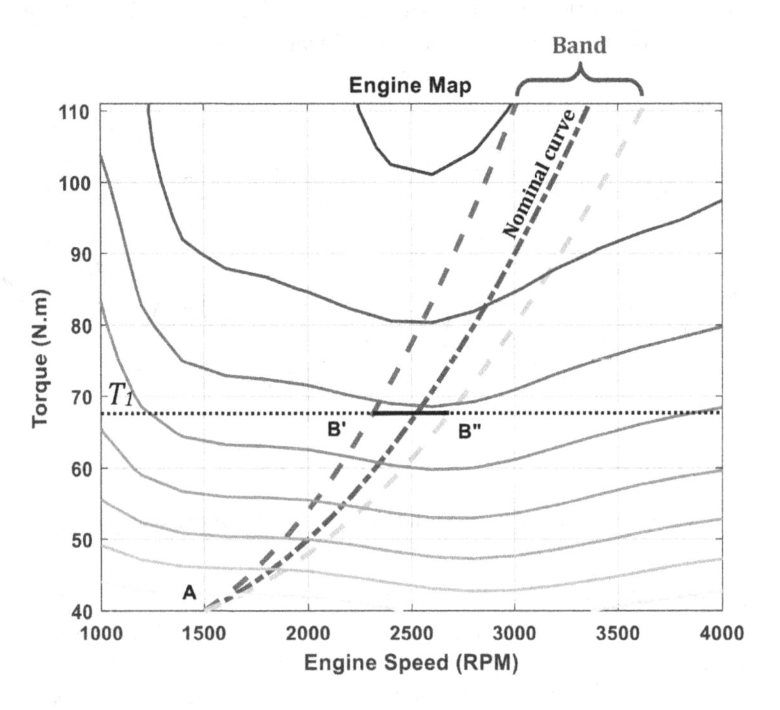

FIGURE 5.5 The uncertainty template of the external load in the fifth gear of the engine [1].

5.4 FORCE-LOAD COMPUTATIONS

In general, for vehicle mass M, equivalent mass of the rotating components M_r, longitudinal acceleration a_x, rolling radius of the tires r, engine torque T_e, numerical ratio of the combined transmission and final drive N_{tf}, combined efficiency of the transmission and final drive η_{tf}, rolling resistance force R_x, aerodynamic drag force D_A, hitch force R_{hx}, vehicle weight W, and road slope θ, vehicle longitudinal dynamic could be stated as follows [3]:

$$(M+M_r)\, a_x = \frac{T_e\, N_{tf}\, \eta_{tf}}{r} - R_x - D_A - R_{hx} - W\, \sin(\theta). \tag{5.2}$$

Since transition is from a steady-state point (A) to another steady-state point (B), T_e could be divided into two parts: T_{e0}, which remains constant during transition, and is balanced with the steady-state load in point A, including R_x, D_{A0}, R_{hx}, and $W\, \sin(\theta)$; and the remaining part ΔT_e. Hence,

$$\begin{aligned} T_e &= T_{e0} + \Delta T_e \\ T_{e0} &= R_x + D_{A0} + R_{hx} + W\, \sin(\theta) \\ (M+M_r)\, a_x &= \frac{\Delta T_e\, N_{tf}\, \eta_{tf}}{r} - \Delta D_A. \end{aligned} \tag{5.3}$$

where $\Delta D_A = D_A - D_{A0}$, and

$$\Delta D_A = \Delta\left(\tfrac{1}{2}\, \rho\, A_D\, C_d\, v_x^2\right). \tag{5.4}$$

Also,

$$v_x = \frac{\omega_e\, r}{N_{tf}} \Rightarrow a_x = \frac{\alpha_e\, r}{N_{tf}}. \tag{5.5}$$

Substituting (5.4) in (5.3), we have

$$a_x = \frac{1}{(M+M_r)} \left\{ \frac{\Delta T_e\, N_{tf}\, \eta_{tf}}{r} - \Delta\left(\tfrac{1}{2}\, \rho\, A_D\, C_d\, v_x^2\right) \right\}. \tag{5.6}$$

Or, by substituting (5.5), we have

$$\alpha_e = \frac{1}{(M+M_r)} \left\{ \Delta T_e\, \eta_{tf} \left(\frac{N_{tf}}{r}\right)^2 - \Delta\left(\tfrac{1}{2}\, \rho\, A_D\, C_d \left(\frac{r}{N_{tf}}\right) \omega_e^2 \right) \right\}. \tag{5.7}$$

Since $\omega_e = \pi/30\; n_e$, where n_e is in rpm, we can write

$$\dot{n}_e = \frac{30/\pi}{(M+M_r)} \left\{ \Delta T_e\, \eta_{tf} \left(\frac{N_{tf}}{r}\right)^2 - \Delta\left(\tfrac{1}{2}\, \rho\, A_D\, C_d \left(\frac{r}{N_{tf}}\right) \left(\pi/30\right)^2\, n_e^2 \right) \right\}. \tag{5.8}$$

In addition, considering $n_e = n_{e0} + \Delta n_e$, we have

$$\begin{aligned}\Delta D_A &= \Delta\left(\tfrac{1}{2}\, \rho\, A_D\, C_d \left(\frac{r}{N_{tf}}\right) \left(\tfrac{\pi}{30}\right)^2 (n_{e0} + \Delta n_e)^2 \right) \\ &= \tfrac{1}{2}\, \rho\, A_D\, C_d \left(\frac{r}{N_{tf}}\right) \left(\tfrac{\pi}{30}\right)^2 \Delta\left((n_{e0} + \Delta n_e)^2 \right) \\ &= \tfrac{1}{2}\, \rho\, A_D\, C_d \left(\frac{r}{N_{tf}}\right) \left(\tfrac{\pi}{30}\right)^2 \left(\Delta n_e^2 + 2\, \Delta n_e\, n_{e0} \right). \end{aligned} \tag{5.9}$$

Consequently

$$\dot{n}_e = \frac{30/\pi}{(M + M_r)}\, \eta_{tf} \left(\frac{N_{tf}}{r}\right)^2 \left\{ \Delta T_e - \Delta D_A \Big/ \left(\eta_{tf} \left(\frac{N_{tf}}{r}\right)^2 \right) \right\}. \tag{5.10}$$

Therefore, in discrete domain

$$n_e(k+1) = n_e(k) + \delta t \cdot \frac{30/\pi}{(M + M_r)}\, \eta_{tf} \left(\frac{N_{tf}}{r}\right)^2 \left\{ \Delta T_e - \Delta D_A \Big/ \left(\eta_{tf} \left(\frac{N_{tf}}{r}\right)^2 \right) \right\}. \tag{5.11}$$

Note that, as mentioned earlier, in hybrid electric vehicles, typically a portion of the total mass $(M + M_r)$ is burdened on the engine as M_{eq}.

5.5 CONTROL

As explained before, to conduct optimal transient control and provide the bandwidth required, explicit nonlinear model predictive control is employed. In general, nonlinear model predictive control (MPC) involves solution of a cost function of a nonlinear model, with terms minimizing transient error (with respect to desired targets) and control cost, subjected to constraints. Here, to attain a convex cost function, the fuel consumed (i.e. injected) over the course of transition is regarded as the control cost.[ix] Note that we sensibly assumed that mid-level controllers work well and engine emissions are properly controlled to conform with the respective driving cycle standard. Accordingly, there is no need to add emission constraints.

The injected fuel (i.e. the control cost) is actually sum of the fuel injected over time increments, which (supposing equal time increments) is linearly proportional to sum of the fuel injection rates over transition increments. Fuel injection rates are, per se, linearly proportional to the air mass flow rates into the cylinders. The latter is in fact the model input. Hence, instead of the injected fuel, sum of the in-cylinder air mass flow rates could be minimized, which, per se, tangibly simplifies the computations.

Now, for the target (T', n'), prediction horizon N_p, and control horizon N_c,[x] the cost function would be

$$\min_{U} J(U, M, N)$$

for which

$$J(U,M,N) = (M - M')^t\, \Omega_T\, (M - M') + (N - N')^t\, \Omega_n\, (N - N') + U^t\, \Omega_u\, U + \Psi \tag{5.12}$$

and

$$\begin{aligned} M &= \begin{bmatrix} T_1 & T_2 & \ldots & \ldots & \ldots & T_{N_p} \end{bmatrix}^t, \quad M' = \begin{bmatrix} T' & T' & \ldots & T' \end{bmatrix}^t \\ N &= \begin{bmatrix} n_1 & n_2 & \ldots & \ldots & \ldots & n_{N_p} \end{bmatrix}^t, \quad N' = \begin{bmatrix} n' & n' & \ldots & n' \end{bmatrix}^t \\ U &= \begin{bmatrix} u_1 & u_2 & \ldots & u_{N_c} & \ldots & u_{N_c} \end{bmatrix}^t. \end{aligned} \tag{5.13}$$

Also, weighting factor matrices are defined as

$$\begin{aligned} \Omega_T &= \frac{1}{\omega_T^{\,2}} \operatorname{Diag}(N_p, N_p), \\ \Omega_n &= \frac{1}{\omega_n^{\,2}} \operatorname{Diag}(N_p, N_p), \\ \Omega_u &= \frac{1}{\omega_u^{\,2}} \operatorname{Diag}(N_{p-1}, N_{p-1}), \end{aligned} \tag{5.14}$$

where M, N, and U represent torque, engine speed, and control command vectors, with corresponding weighting matrices Ω_T, Ω_n, and Ω_u, respectively. Furthermore, ω_T, ω_n, ω_u denote normalizing factors for torque, engine speed, and air mass flow rate, respectively. $\operatorname{Diag}(N_p, N_p)$ means a diagonal matrix with N_p rows and N_p columns. The diagonal arrays for Diag matrices in Ω_T and Ω_n had better be chosen as a (linear or nonlinear) sequence of squared integers from 1 to p^2, in order to magnify the importance of later errors (i.e. the later arrays of the error matrices $(M - M')$ and $(N - N')$, for better convergence to the desired targets (i.e. M' and N').

The penalty matrix, Ψ, represents the constraints. Note that instead of separately defining a constraint, it is incorporated as a penalty term into the cost function itself, so as to accelerate the computations. They can include different constraints, from system variable constraints (e.g. throttle angle constraints, as they cannot take values larger than 90°, or values smaller than idling throttle angle), to state constraints (i.e. constraints for the produced torque and engine speed in each step). Large values are assigned to penalty terms so as to inhibit convergence to undesired results, and/or to accelerate the convergence procedure.

An example of optimal transition curves is displayed in Figure 5.6. The respective control command is indicated in Figure 5.7. The starting point has also been selected as an example.[xi]

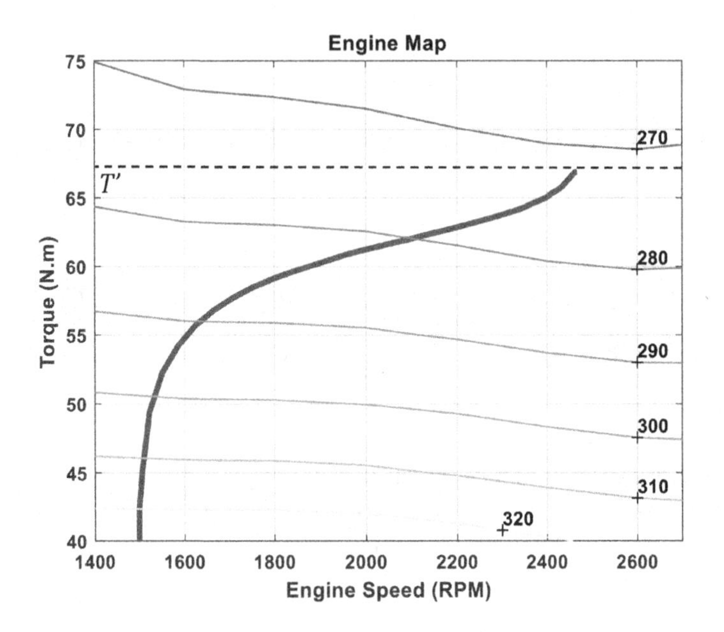

FIGURE 5.6 An example of the optimal transition curve [1].

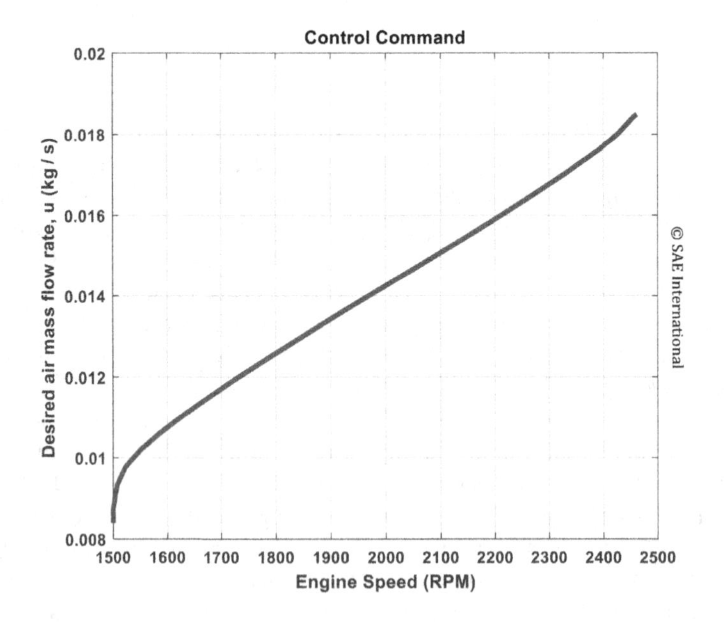

FIGURE 5.7 Control command (U) for the optimal transition curve [1].

The resulting off-line curve above could be stored in the high-level control layer.[xii] The curves could also be converted to polynomials, whose coefficients are stored as look-up tables or other efficient black-box identifiers.

As explained before, the estimated external load carries uncertainty. This uncertainty causes the control system (with the above off-line controller) to converge to points different from the desired targets. This is not an issue as for ordinary vehicles (as it can be compensated by the driver further depressing the gas pedal), yet matters when it comes to hybrid vehicles, since it could disturb the strategy of energy management system (EMS).

Basically, it is "Receding Horizon" that provides robustness for MPC. Nevertheless, this involves solving the problem for a polytope around every single point between starting and target points, which, apart from cumbersome computations, is impossible for ECU in terms of memory. Furthermore, this demands a load observer able to precisely estimate the instantaneous load on-line, which is practically unattainable. Consequently, Receding Horizon cannot be performed, resulting in robustness problems.

The recourse could be combining the off-line controller with a low-gain PI controller (as carried out in our paper [4]). Another option is to apply (what we call) one-step receding horizon. That is, for the system with unknown uncertainty, first a certain number of the beginning control commands derived for the nominal system (e.g. the beginning q arrays of the vector U derived for the system with nominal load) are applied. Afterward, based on the resulting engine operating point, an estimation of the actual load could be attained. In other words, for the certain beginning q arrays of the nominal control commands exerted to the system (with unknown external loads and uncertainties), the resulting point (T_{k+1}, n_{k+1}) would be different from that with the nominal external load, and in a sense, can indicate the actual external load. Now, if the control system behavior (i.e. curves) for different possible external loads (and with the same beginning q arrays of the nominal control commands exerted to the system) have been derived and stored beforehand, the resulting point (T_{k+1}, n_{k+1}) can readily reveal which load, after k steps, would make the system converge to that point (Figure 5.8). Accordingly, the remaining of the control command sequences [i.e. $U(q+1\text{: end})$], required to converge the system to the desired torque T', could have been solved off-line [using the optimization formula (5.11)] and stored for every different external loads within the uncertainty template. During on-line control, these stored commands would be applied to direct the system to the demanded torque. Needless to say, although the target torque remains the same as that of the nominal load (i.e. T'), the target engine speed would be different (i.e. the intersection of the new load curve and the original demanded torque line T'). Figure 5.8 illustrates some examples of the methodology above for different external loads within the respective uncertainty template.

It shall be noted that the engine torque estimation for allocating the point (T_{k+1}, n_{k+1}) in this strategy, is just like in EMS; it could be readily estimated using a simple combustion model as torque observer. Since all the inputs to combustion model are known, and the combustion subsystem is static (Section 2.4.5), the produced torque can be easily estimated.

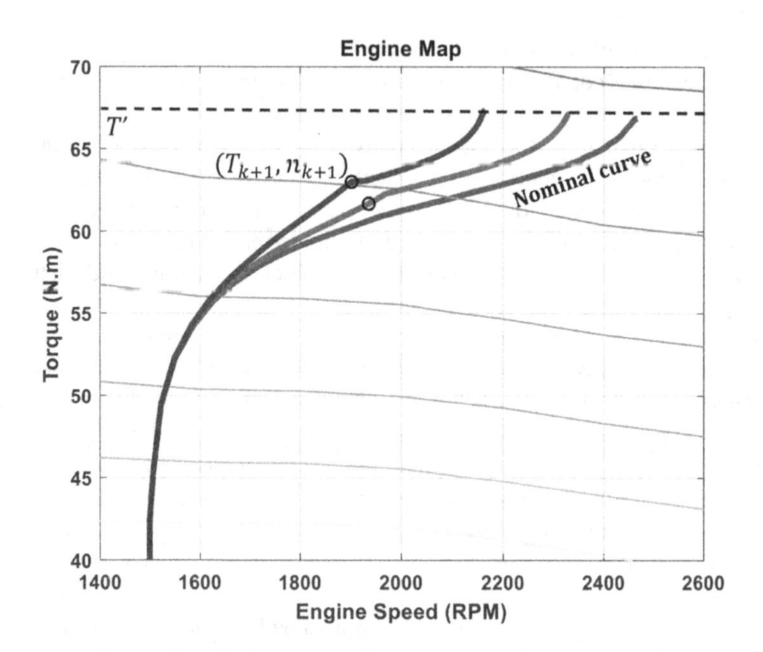

FIGURE 5.8 Examples of one-step receding horizon for different loads within the uncertainty template. As can be seen, the resulting points (T_{k+1}, n_{k+1}) are different from that of the nominal external load, indicating which pre-solved curve would suit the existing unknown load [1].

It is worth reiterating that with regard to hybrid electric vehicles, the number of operating points tend to be limited, and the above control strategy does not cause any impediment in terms of the memory required.

5.6 SUMMARY

In this chapter, engine high-level torque control was examined, and a new idea, useful particularly for hybrid electric vehicles was presented. As stated, transients are of great importance in terms of fuel consumption, especially for hybrid vehicles as they are frequently started off and loaded up to demanded torques. To provide the bandwidth and the agility required, pure feed-back control was avoided, and a semi-feed-forward controller was incorporated into high-level layer of the hierarchical control structure. Using the methodology explained, optimal transition curves could be derived, and stored in the ECU, so as to minimize transient fuel consumption. In order to address uncertainties in external loads applied and to provide robustness, a new approach named one-step receding horizon model predictive control was proposed. Since hybrid vehicles typically tend to deal with a limited number of operating points (compared to an ordinary vehicle), the approaches proposed do not cause any obstacle in terms of the memory required.

For implementation in a hybrid electric vehicle, a comprehensive thorough study concerning desired torques, loads, and uncertainty templates need to be conducted.

NOTES

i. At most, between 0.2–0.02 s, depending on the engine speed.
ii. Also recall that a large portion of engine operation (particularly) in parallel hybrid vehicles occur outside of sweet spot.
iii. As a matter of fact, there are studies in the literature aiming to improve transient behavior of the control system by considering the low-level or mid-level controllers. However, as noted, the influence of the high-level controller is more substantial in transients, and this is where we will construct our controller.
iv. "*T*" stands for torque, and "*n*" denotes the engine speed.
v. As for hybrid vehicles, a part of external load could come from charging the battery.
vi. Note that this approach is more general than sweet spot.
vii. For a more general approach, ambient air circumstances could also be regarded as inputs. The produced emissions could additionally be considered as outputs.
viii. It shall be noted that in a hybrid electric vehicle, with electric motor under load, only a portion of the vehicle mass would be imposed to the engine. However, this does not change the optimal curve derived in the end.
ix. The pollutants emitted over the course of transition could also be incorporated into the cost function. However, this could make the cost function non-convex, immensely complicating the optimization.
x. A good choice for *Nc* is $N_c = N_p - 1$.
xi. Normally, the starting point is probably around idle. Nevertheless, here, it is meant to demonstrate that even at the presence of extra non-transient load, the approach is valid.
xii. Control commands could be allocated based on the feed-back from the rpm sensor (which results in a semi-feed-forward controller), or the sequence of arrays in vector U stored.

REFERENCES

1. Shamekhi, A.-M. and A.H. Shamekhi, *Engine high-level nonlinear model predictive torque control with enhanced application for hybrid vehicles.* SAE International Journal of Engines, 2022. **15**(2): p. 283–295.
2. Guzzella, L. and C. Onder, *Introduction to Modeling and Control of Internal Combustion Engine Systems.* 2nd ed. 2010, Berlin, Germany: Springer.
3. Gillespie, T.D., *Fundamentals of Vehicle Dynamics.* 1992, Warrendale, PN, USA: Society of Automotive Engineer.
4. Shamekhi, A.-M., A. Taghavipour, and A.H. Shamekhi, *Engine idle speed control using nonlinear multiparametric model predictive control.* Optimal Control: Applications and Methods, 2020. **41**(3): p. 960–979.

Appendix A
A Short Review of Neural Networks Design

A.1 INTRODUCTION

Neural network is a black-box model based on learning if-then rules. Input–output mapping in neural networks is inspired by the incredibly sophisticated process of learning in the human brain. When a little child strives to master climbing stairs, for example, he or she is actually trying to *learn* the manner and the extent to which they should move their hands and feet through numerous trial-and-errors. In fact, at the outset, when they frequently fall down, the command signals in their brains (that control the movement of body parts) are not still regulated. As time passes, due to these trial and errors, and of course, the innate tendency of a human being toward perfection, the child's brain masters how much every signal should be. It is to be noted that the human brain accomplishes this procedure not by solving the physical equations governing the body (which are extremely complex and of enormous degrees of freedom) but by utilizing input/output mapping (or the if-then rules). Obviously, the cost necessary for this learning is exposure to enough inputs (or the child's numerous trial and errors) and a kind of supervision or a reward/penalty system (or here, the innate tendency toward perfection). Quite a simplified version of this method is employed as Artificial Neural Networks in modeling (i.e. system identification), clustering, and pattern recognition. The schematic of a neural network is indicated in Figure A.1 [1].

Neural networks have a remarkable capability of black-box modeling of nonlinear phenomena. Because of only dealing with input/output data, they are not entangled with un-modeled dynamics (unlike white-box models). What distinguishes neural networks from other black-box practices is the fact that they are intelligent to some extent, and thus, require lesser empirical data. They do much better at interpolation than extrapolation.

However, as a matter of fact, as the complexity of the system to be modeled increases, the ability of a network (to accurately model and estimate it) declines. Neural networks are primitive imitations of a human brain. Therefore, whatever is difficult for the human brain to deal with, would be definitely difficult for neural networks to handle, as well.

Of all neural networks, Multi-Layer Perceptron (MLP) networks are probably the most prevalent type. In general, these networks are of two kinds: static and dynamic. This appendix is to examine principles and practices in the optimal designing of MLP neural networks.

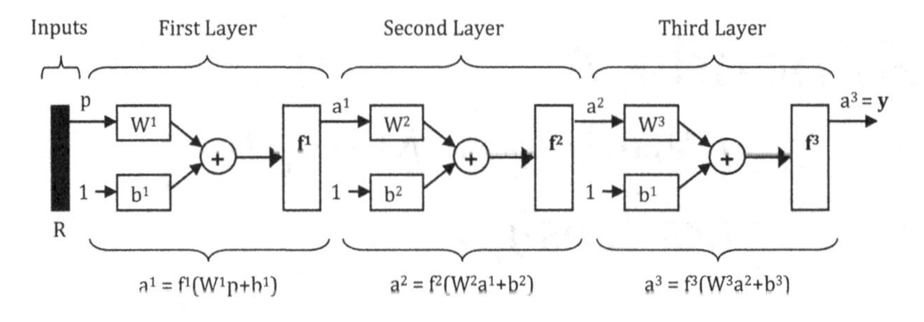

FIGURE A.1 The schematic of a three-layer neural network [1].

In what follows, rudimentary definitions and descriptions are ignored, and only major challenges are mentioned. A short glance at the dynamic MLP networks is provided at the end, in Section A.12.

A.2 TABLE OF PATTERNS

The table of patterns (or data) is divided into three sets: training data set, cross-validation data set, and test data set. All the algorithms of training neural networks aim to minimize the training set error. After training, the test data set error will be the criterion for assessing the generalization capability of the network. Typically, as for a trained network, the mean squared error (MSE) of the test set is most certainly higher than that of the training set (Figure A.2). An ideal measure is achieved when the number of patterns in the test data set approaches infinity. The training error[i] is typically less than the accurate error of a network. As illustrated in Figure A.2, when the number of samples (or patterns) approaches infinity, the two training and test errors converge to the accurate value [2].

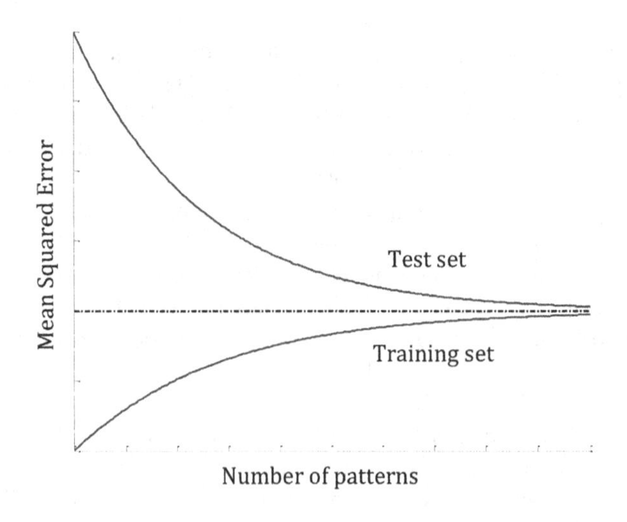

FIGURE A.2 The influence of pattern number on training and test errors [2].

It might seem that the network's accuracy decreases with the increase in the number of samples (as the training error rises). But in fact, the test error (which is the genuine criterion of a network's performance) declines as the number of samples grows. Hence, in terms of network accuracy, the higher the number of patterns, the better.

Obviously, it is not possible to provide a table containing an infinite number of samples. As a matter of fact, samples are usually derived from experiments, and thus, are finite. However, the number of these data must be large enough. Figure A.3 illustrates the outputs of two networks with identical structures each of which is trained by a different pattern table. Solid-line represents the genuine result, and dashed lines stand for the outputs of the networks. The picture on the right belongs to a neural network trained with 20 samples, but the left one has been trained with only four samples. As can obviously be seen, an excessively small number of data can result in a huge error [2].

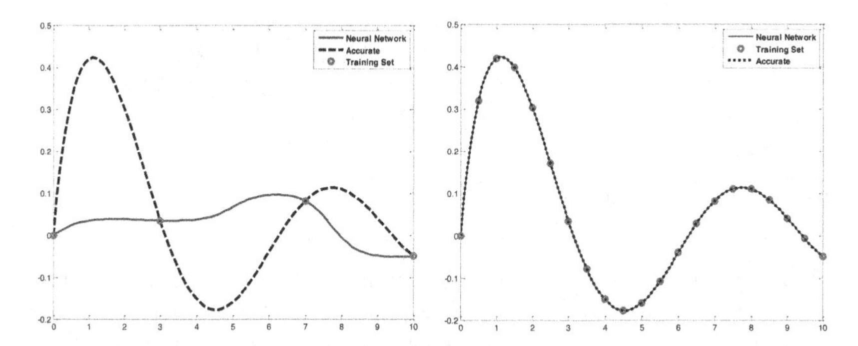

FIGURE A.3 The influence of pattern numbers on the generalization capability of neural networks. As can be seen, training a network based on a too small training set (left) causes weak generalization, as opposed to a large training set (right).

Moreover, the samples need to include all different operating regions of the system. In critical areas, for example, when the behavior of the system contains drastic changes, more data (i.e. higher density of data acquisition) is required. As for the case represented in Figure A.3, a higher density of samples is needed in the first swing than in the second one.

It is of great use to carry out input-output processing. It involves pre-processing for inputs and post-processing for outputs. For tangent-sigmoid transfer functions to work well, the inputs need to be within a limited range of around zero. MATLAB software automatically uses some preprocessing and post-processing functions that have a great impact on the efficiency of the networks to be trained. For example, the function "mapstd" normalizes the inputs and desired outputs (or targets) such that they have zero mean and a standard deviation of one. The function "processpca" utilizes the principal-component analysis to reduce the dimension of correlated input vectors (refer to MATLAB Help for more information).

A.3 PERFORMANCE SURFACES OF NEURAL NETWORKS AND INITIAL WEIGHT VALUES

A neural network, at first, contains a number of free parameters (or weights) which shall be gradually adjusted over a training process. The performance (or error) function of a network forms an $n+1$-dimensional surface,[ii] which is called the performance surface. A simple three-dimensional example of these surfaces is shown in Figure A.4.

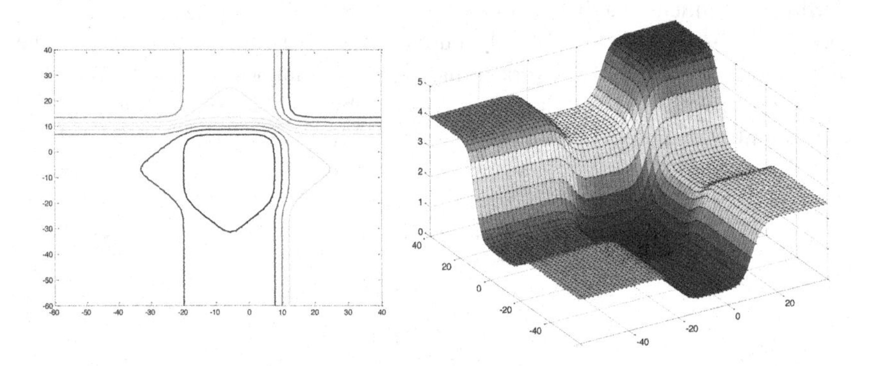

FIGURE A.4 A schematic of a performance surface.

Starting from assumed initial weights, all the training procedures try to reach the global minimum of the surface (i.e. the global minimum of the network error). Meanwhile, the great dilemma is not being stuck in local minima but reaching of the real global one.

As regards initial weights, small values must be chosen. Since training methods proceed with derivative values of error function at each step, zero value (which is saddle point) or points too far from the origin (that have extremely small derivatives due to the nature of tangent-sigmoid functions) are not suitable for initial weights and impede the process of training [1]. The Neural Networks toolbox of MATLAB software automatically takes these considerations into account. However, it is to be noted that, in the end, the training procedure might not converge to the global minimum and may get stuck in one of the local minima. Therefore, when training a network, enough different initial weights must be examined to make sure of convergence to the global minimum.

A.4 NUMBER OF HIDDEN NEURONS

As for an MLP neural network, after providing a comprehensive pattern table, the structure of the network shall be determined. According to the Universal Approximation Theorem, a single-hidden-layer network, with a limited number of sigmoid transfer functions in the hidden layer and linear transfer functions in the output layer, is capable of approximating any continuous function on compact subsets of R^n [1]. Yet, two hidden-layer networks yield deeper structures, with a higher level of feature extraction.

As for the number of hidden neurons, more hidden neurons mean a more complex performance surface, and accordingly, a greater capability of approximation. On the other hand, if the complexity of the performance surface overly rises, the probability of being stuck in the local minimum (when training the network) builds up. Therefore, as a general rule, a neural network should not be unreasonably complicated.

The network loss function could be defined as the expectation of the squared error; namely, $E\{(e)^2\}$, which is:

$$E\{(e)^2\} = E\{(y-\hat{y})^2\}, \tag{A.1}$$

where y is the measured output (with noise), and $\hat{y}$ is the estimated output by a neural network. Considering noise n, we have:

$$E\{(e)^2\} = E\{((y_u+n)-\hat{y})^2\} = E\{(y_u-\hat{y})^2\} + E\{n^2\}, \tag{A.2}$$

in which y_u is the exact output (without noise).

The term $y_u - \hat{y}$ is the model error and can be decomposed as follows [3]:

$$E\{(y_u-\hat{y})^2\} = [y_u - E\{\hat{y}\}]^2 + E\{[\hat{y} - E\{\hat{y}\}]^2\}. \tag{A.3}$$

The above equation can be rewritten in the following terms:

$$(\text{Model Error})^2 = (\text{Bias Error})^2 + \text{Variance Error}. \tag{A.4}$$

Bias error is a part of the model error which pertains to limitations in the model's *flexibility* to represent (or resemble) the original system. In practice, the majority of processes are extremely nonlinear and convoluted, and no model can *exactly* represent them. Therefore, this error is entirely an imperfection related to the model structure. As the number of hidden neurons increase (i.e. as the network's parameters rise), the model becomes more flexible, and thus, bias error declines (Figure A.5). However, typically, a non-linear system cannot be modeled without the bias error.

Variance error is a part of model error that results from the deviation of estimated *parameters* from their optimal values. Since model parameters, in practice, are calculated from a set containing a *limited* number of *noisy* data, they deviate from accurate values. In other words, variance error represents a part of model error which is due to uncertainty in calculated parameters. Due to noise and the resulting uncertainty, the smaller the number of (free) parameters, the more accurate they are estimated. Should the number of samples equal the number of

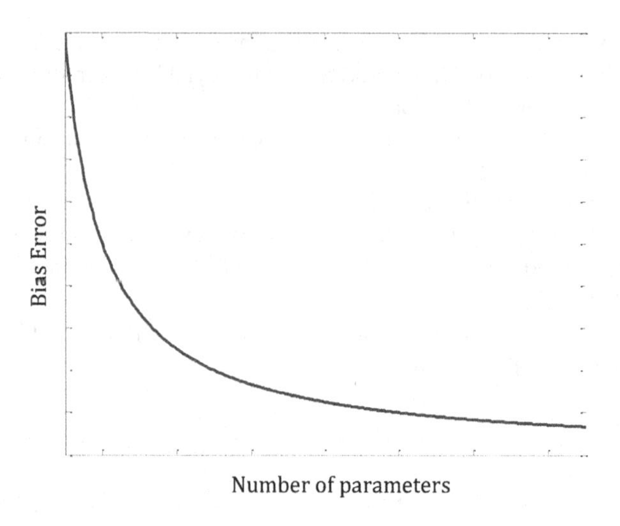

FIGURE A.5 The influence of model parameters on bias error.

parameters, the variance error reaches its peak. Hence, in contrast to bias error, *as the number of hidden neurons increases, variance error grows* [4]. A neural network with more-than-enough parameters is called over-determined.

Variance error almost linearly grows with increasing network parameters (Figure A.6), and declines with increase in training data. In addition, noise has a major impact on this error and exacerbates it. Namely [3]

$$\text{Variance Error} \approx \text{Noise} \cdot \frac{\text{Number of Free Parameters}}{\text{Number of Training Data}}. \tag{A.5}$$

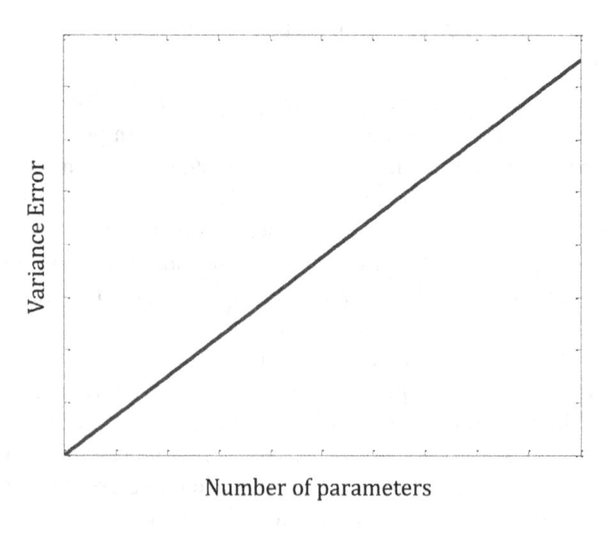

FIGURE A.6 The influence of model parameters on variance error.

Recall that total model error is sum of bias error and variance error. As noted above, increasing hidden neurons, on one hand, reduces bias error, and on the other hand, raises variance error. This is called the *Bias/Variance Dilemma*. As can be seen in Figure A.7, there exists an optimal value for total model error and the number of hidden neurons. Accordingly, the primary way is to find the optimal number of hidden neurons is to gradually raise the number of hidden neurons until the optimal number is obtained [4]. Besides, the samples ought to be as less noisy as possible.

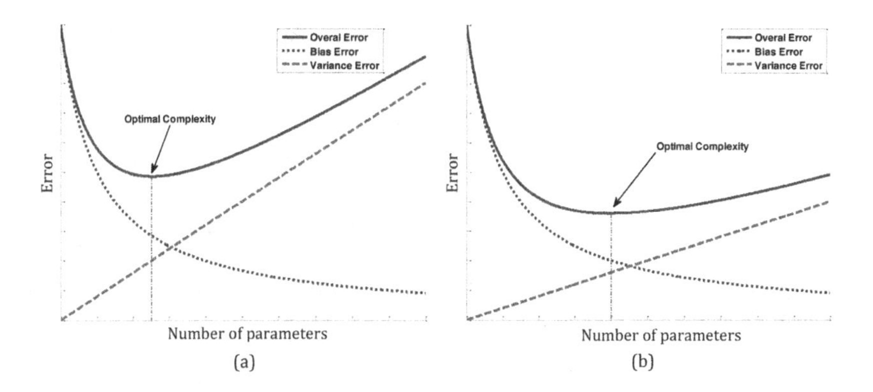

FIGURE A.7 The compromise between bias and variance errors, and the effect of noise on the compromise. The data based on which the network (b) is trained had been less noisy than those of (a). Therefore, the number of optimal parameters, and consequently, the optimal error is less in (b) [4].

A.5 THE NUMBER OF ITERATIONS IN TRAINING PROCESS

As will be explained in Section A.6, training algorithms attempt to modify synaptic weights in a gradual iterating process. The primary function of a network, as mentioned before, is generalization. That is, a network is supposed to predict other points present in the interval almost as accurately as training points. These other points are represented by test data set.

When training a network, training error declines as the number of epochs (i.e. the number of iterations) increase. However, training data are normally derived from experiments, and thus, are noisy. A network must conduct something of a curve fitting through these points. This n-dimensional curve does not need to pass exactly through all these *noisy* training points, but rather, it should minimize the generalization error. Over-training makes the network *memorize* all these noisy data rather than fitting a suitable curve through them [4]. The consequence would be a weak generalization (Figure A.8). Over-training accompanied with over-determining are the primary manifestations of over-fitting.

To avoid over-training, cross-validation data set (which is not used for training, just as test set) is employed as the stopping criterion of training a network; that

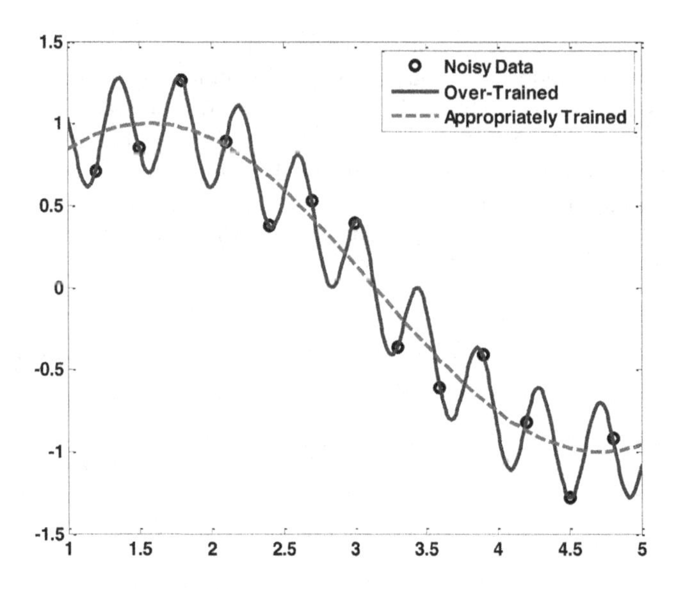

FIGURE A.8 A schematic diagram representing the effect of over-training on the generalization capability of a network when training a noisy set of data.

is, when training a network using training data, once the cross-validation error rises in the last 10 epochs, for instance, the training stops. This is called "early stopping" and implies the fact that training a network must be carried on up to point which the network's generalization ability does not decline (Figure A.9).

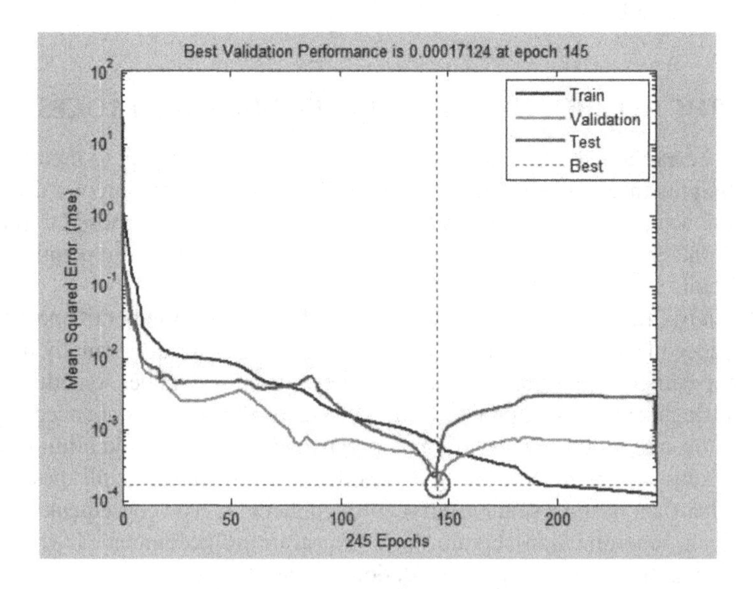

FIGURE A.9 Early stopping based on cross-validation error. As can be seen, at the best validation performance, training error keeps falling while cross-validation and test errors rise.

In the end, the test set will be the criterion for assessing a network generalization. Obviously, to train a network properly, appropriate training, cross-validation, and test sets (all thorough representatives of system behavior) need to be employed with suitable adjustment of early stopping.

A.6 TRAINING METHODS

Generally, there are three methods to reduce the error function of a network in iterating practices. All these methods involve gradually optimizing network weights. They are as follows [4]:

a. **Steepest Descent Algorithm:** In this method, updating weights, in each step, is carried out towards the steepest descent direction, namely, the negative of the gradient vector:

$$w(n+1) = w(n) - \eta \nabla E(w) \tag{A.6}$$

where $w(n)$ is the weight vector at the nth iteration, $E(w)$ stands for error function, evaluated at $w(n)$, and η represents learning rate.

This method is too slow and the training performance is excessively affected by the size of η.

b. **Newton Algorithm:** The major idea of this approach is to minimize a second-order approximation of the cost function around $w(n)$ in each step. As for the cost function, we have

$$\Delta E(w(n)) \approx g^T(n)\Delta w(n) + \frac{1}{2}\Delta w^T(n) H(n) \Delta w(n) \tag{A.7}$$

where $g(x)$ is the gradient vector of error function, $H(x)$ represents the Hessian matrix (containing second-order derivatives of the error function with respect to weights). Deriving the above relation and equalizing it to zero yields

$$w(n+1) = w(n) - H^{-1}(n) g(n). \tag{A.8}$$

This algorithm converges asymptotically, without fluctuations of the gradient descent method. Nonetheless, for the above relation to have a solution, the Hessian matrix needs to be a positive definite in all iterations, which is mostly not the case.

c. **Gauss–Newton method:** Using Jacobian matrix, this algorithm presents an approximation of Hessian matrix as $2J^T(n)J(n)$, and finally reaches the below relation:

$$w(n+1) = w(n) - \left(J^T(n)J(n)\right)^{-1} J^T(n) e(n). \tag{A.9}$$

However, the value of $J^T(n)J(n)$ shall be nonsingular. To do so, it is modified by the diagonal matrix δI, in which, δ is a positive small value:

$$w(n+1) = w(n) - \left(J^T(n)J(n) + \delta I\right)^{-1} J^T(n)e(n). \tag{A.10}$$

The aforementioned algorithms are the basis for various methods of training neural networks. The methods of "Steepest Descent" and "Steepest Descent with Momentum" are two examples derived from the first algorithm, and slowest methods of training. Sluggishness of a method not only means longer solving time, but it also, in practice, causes an increase in the minimum error attainable for a network [5].

With the passage of time, some other methods were presented which were ten to a hundred times faster. These approaches, which all exploit batch mode of training, are classified into two different categories. The first category utilizes heuristic techniques derived from the operation of steepest descent method. The two approaches of "Variable Learning Rate Back-propagation" and "Resilient Back-propagation" are of most importance in this kind.

The second type uses numerical optimization techniques. In this regard, "Levenberg-Marquardt" is the most famous and the most commonly used one, employing Gauss–Newton algorithm. As previously mentioned, Newton and Gauss–Newton methods minimize second-order estimations of error function, and are of great speed and precision. However, due to calculation and saving Hessian matrix (namely, second-order derivatives of all outputs with respect to all the weights), they demand huge computer footprint. This restricts the competency of these methods for large networks. On the other hand, "Conjugate Gradient Method" uses conjugate vectors rather than Hessian matrix. While slower and less accurate than Levenberg-Marquardt method, the conjugate gradient method is more suitable for large networks [5].

All in all, the algorithms of the resilient method, conjugate gradient method, and Levenberg-Marquardt may be considered the best ones. As for networks with small or medium sizes, Levenberg-Marquardt method is adequately fast with best precision.

Using Bayesian Framework, the neural networks toolbox in MATLAB, has presented "Bayesian Regularization" algorithm (trainbr). As mentioned before, to attain higher generalization, hidden neurons need to be gradually increased so that the optimum number of neurons is achieved. Bayesian Regularization is capable of automatically calculating the optimum number of hidden links. In this approach, hidden parameters are gradually increased such that the cross-validation error does not grow. Obviously, for this method to be effective, "early stopping" needs to be restrained to let a network train until convergence.

Other methods employ the cross-validation set merely for early stopping and cease training when the cross-validation error rises in several successive epochs. On the other hand, Bayesian regularization utilizes the cross-validation set not only to inhibit overtraining but also to adjust the number of hidden links, and thus, to impede over-determining.

Moreover, as early stopping is curbed, the network has the chance to get out of local minima and keep training until converging to the global minimum. In other words, in Bayesian regularization, a network has the chance to be completely trained while not being over-trained.

Accordingly, when compared to Levenberg-Marquardt, although Bayesian regulation might have higher training errors, it may reach less test errors, signifying better generalizations.

In the end, it shall be noted that test error performance should have a logical descending trend just as the cross-validation error. For example, the trend depicted in Figure A.10 is not descent.

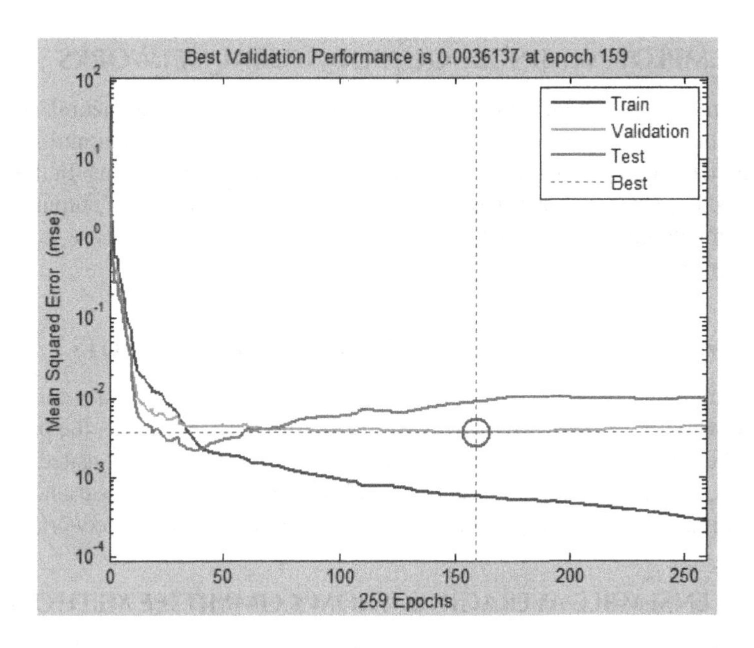

FIGURE A.10 An *inappropriate* trend of test error indicating an unacceptable training.

A.7 PARSIMONY

According to the parsimony principle, of two neural networks with the same level of error, the simpler network with less hidden neurons and simpler structure is preferable [3].

More complicated networks (with more neurons) have more convoluted performance surfaces. Therefore, when training, the possibility of being stuck in local minima rises and the network becomes less efficient. Furthermore, for a given number of training patterns, the variance error grows as the number of free parameters increases (i.e. as the number of neuron increases). This inevitably causes larger levels of error.

It can also be inferred that the simpler a task, the more preferable the erected network would perform. More complex tasks require more convoluted networks, meaningless favorable network performances. Consequently, breaking down a complicated task into simpler ones and training networks for each one may give much better overall performance.

As mentioned before, neural networks are immensely simplified versions of the human brain. Hence, whatever is difficult for the human brain can be difficult for neural networks as well. Therefore, just like in the case of human brain, simplifying tasks burdened on neural networks can be an effective recourse to improve their performance. In what follows, some examples of network/task simplification are presented.

A.8 EMPLOYING SINGLE-OUTPUT NEURAL NETWORKS

As for modeling a multi-output system, instead of training one neural network containing multi outputs, it would be much more effective to train multi single-output networks each of which account for one of the system outputs. In this way, the task burdened on each network is simplified, and thus, each output is predicted much more accurately (just like applied in [6]). This, in a sense, coincides with parsimony principle.

A.9 MINIMIZING THE NUMBER OF NETWORK INPUTS

Including inputs of negligible impact in a neural network can uselessly complicate the network structure and the relevant performance surface; hence, the probability of being stuck in local minima escalates (an example is investigated in [7]). Besides, raising the number of network parameters increases the variance error. Accordingly, care should be taken when selecting the inputs of a network.

A.10 ENSEMBLE AVERAGING (FROM COMMITTEE METHODS)

The committee method has been proposed in order to overcome bias/variance dilemma. As mentioned, bias/variance dilemma restricts the accuracy attainable for neural networks. This method can be divided into two general categories: ensemble averaging and boosting. In ensemble averaging, inputs are identically given to a bunch of over-determined network and the average of outputs is calculated (Figure A.11).

Due to over-determination, the bias error of these networks is less than that of the optimally trained single network. However, the variance error rises owing to over-determination. It is mathematically proven that the connected structure, shown in Figure A.11, would be of less variance error than all the members, while the bias error remains unchanged [4]. In this way, the bias/variance dilemma is conquered.

In other words, instead of choosing only one optimal network from many trained networks and discarding the rest, all the trained networks can be exploited

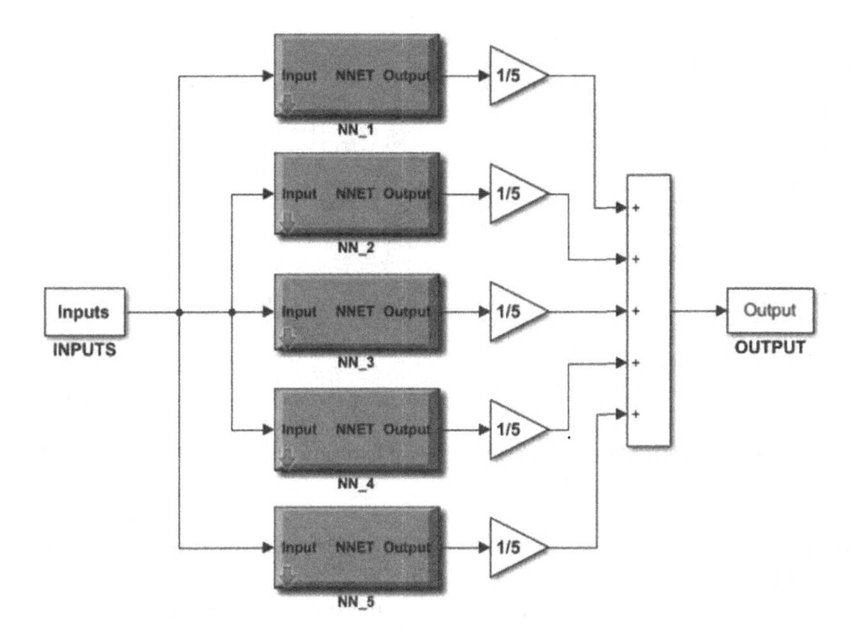

FIGURE A.11 An Example of Ensemble Averaging structure, for 5 networks.

in a committee structure. Interestingly enough, the result may be overwhelmingly better. It is reaffirmed that for this method to work more effectively, the members need to be slightly over-fit.

The averaging gain for networks (i.e. one-fifth in Figure A.11) does not have to be the same for all the members and can be optimally calculated for each one. The method is explained by Hashem [8].

It is to be noted that the variance error progressively declines as the number of committee members increases. However, this action builds up the response time of the model.

A.11 IMPROVED PARTITIONING METHOD

Sometimes, when a huge operating area is to be taught to a network, it is observed that the network trained does not show a favorable performance on all the area. For instance, the MSEs as to outputs of large values are relatively low, while small desired outputs are presented with large errors (Figure A.12).

This is primarily due to two reasons: one is the huge range of outputs. In this case, the network cannot learn both too small and too large values. Thus, due to the higher impact of larger values on the overall error, the network tends to learn them better than the smaller ones.

The other reason is substantial change in output pattern; in other words, when system behavior shows drastic changes as one input varies. For example, the output pattern as for inputs higher than a specific value is totally different from those

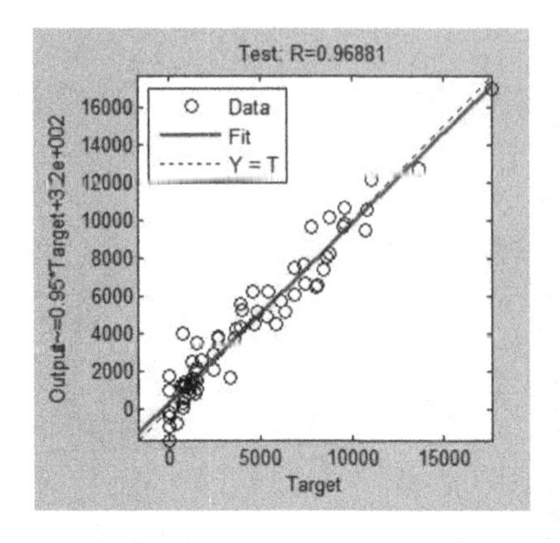

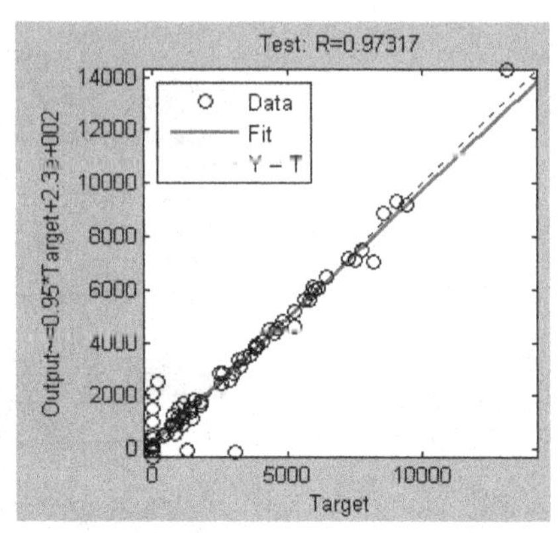

FIGURE A.12 The unfavorable performance of two networks as for smaller targets.

for lower values. In this case, it will be too difficult for a network to learn all those essentially distinct behaviors (or regimes).

One way to surmount this issue is to employ two (or more) networks to learn different regimes of the system. That is, each regime is allocated to one network (or one committee of networks). For instance, one network learns large amounts and one other learns lower ones. This partitioning concept, again, ties in with simplifying the tasks burdened on neural networks, and was the basis for introduction of ANFIS [9], and local model trees, e.g. LOLIMOT [10].

However, this approach involves partitioning the training set corresponding to the different regimes, and therefore, severely reducing the number of training data available for each network to learn the relevant regime. As indicated in [7], this method might not be so effective since shortening the training set has an adverse effect on network performance and raises the variance error. Furthermore, in this approach, the erected networks might be weak at predicting data adjacent to partitioning borders.

To overcome the above problem, we add a part of patterns adjacent to partitioning borders to both neighboring partitions such that both partitions partly overlap (Figure A.13). In this way, not only will the reduction in pattern numbers for each partition be (partly, to say the least) compensated, but the trained networks will also predict the border regions much better. In this way, the MSE can be significantly reduced (up to 75% reduction, compared to the optimal individual network in the authors' previous article in 2015). We have called this method "Improved Partitioning."

Note that instead of the method above, ANFIS, LOLIMOT, HOLIMOT, etc., could be employed. However, for the sake of more flexibility in utilizing ensemble averaging, we prefer to use improved partitioning. In fact, employing the combination of improved partitioning and ensemble averaging can remarkably reduce MSE (up to 90% reduction of MSE, compared to the optimal individual network in the authors previous article in 2015 [11]).

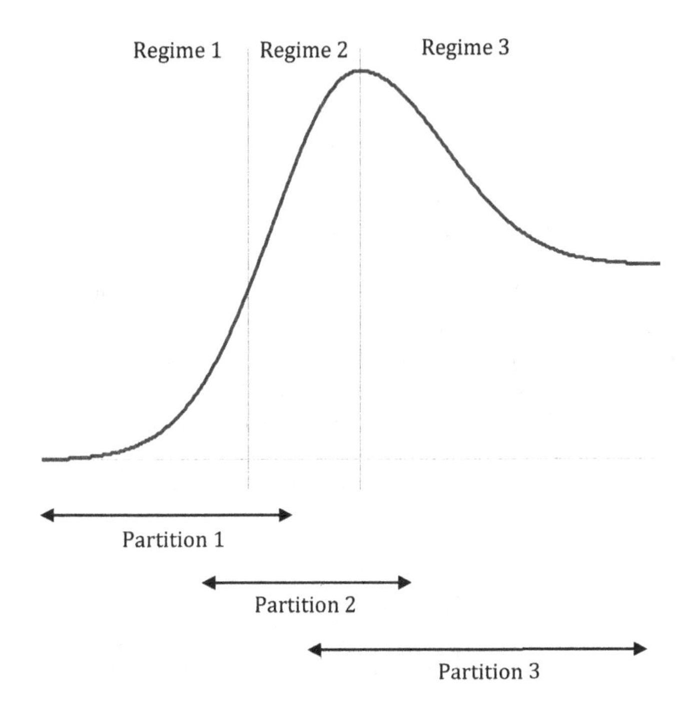

FIGURE A.13 Extended, overlapped partitioning; each extended partition is allocated to one network (or one committee of networks).

A.12 DYNAMIC NEURAL NETWORKS

In general, for a neural network to have memory, time delay is needed. When the delays incorporated are feed-forward, then the network will be of short-term memory. On the other hand, delays contained in feed-back links bring long-term memory to the network.

Modeling some systems, like speech recognition, involves memorizing (only) some steps behind. In these cases, feed-forward delay structures can be used. However, when modeling a dynamic system, feed-back time-delay neural networks, or so-called "Recurrent Neural Networks (RNN)" must be applied. Major issues in recurrent networks are stability and training. It is to be noted that incorporation of a feed-back delay in a network can cause instability. Therefore, after designing a recurrent network, its stability needs to be established using Lyapunov's first theorem[iii] [4].

Due to delays, training algorithms of dynamic networks are far more demanding than static neural networks. Furthermore, performance surfaces of recurrent networks are much more complicated, and, thus, the probability of being trapped in local minima when training is much further. Hence, dynamic networks demand training over and over with different initial weights [5].

Feed-back incorporation in a network can be implemented locally (i.e. with regard to only some of the nodes) or globally.

A dynamic system in discrete state space is defined as follows:

$$x(n+1) = f(x(n), u(n))$$
$$y(n) = Cx(n). \tag{A.11}$$

In the above relation, $\boldsymbol{u}$, $\boldsymbol{x}$, and $\boldsymbol{y}$ represent input vector, state vector, and output vector, respectively. The state vector in the next time step (i.e. $\mathbf{x}(n+1)$) is defined as the nonlinear function f(.) of the input and the state vectors at the current time step (i.e. $\mathbf{u}(n)$ and $\mathbf{x}(n)$). Moreover, system output is achieved from a linear operation on state vector (namely, C $\mathbf{x}(n)$) at discrete time n. Here, the idea of feed-back time-delay neural networks is derived. That is, nonlinear functions of hidden neurons can be employed as nonlinear function f, and the linear function in the output layer can be assumed as the constant matrix C. The inputs of hidden neurons are network inputs (i.e. $\mathbf{u}(n)$) and fed-back delayed states (i.e. $\mathbf{x}(n)$). This structure is called "Elman Network" or state-space model (Figures A.14 and A.15) [4].

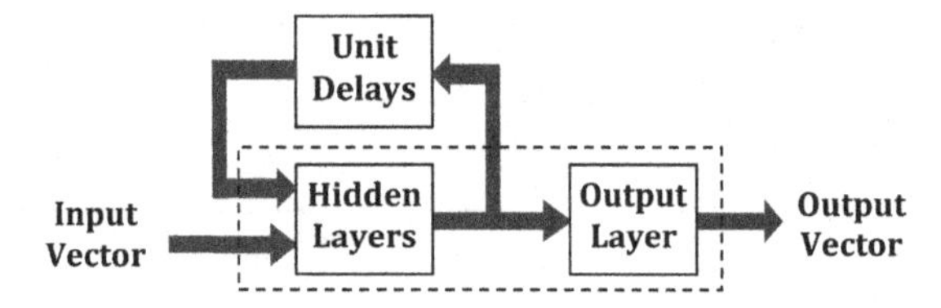

FIGURE A.14 Schematic of a state-space model [4].

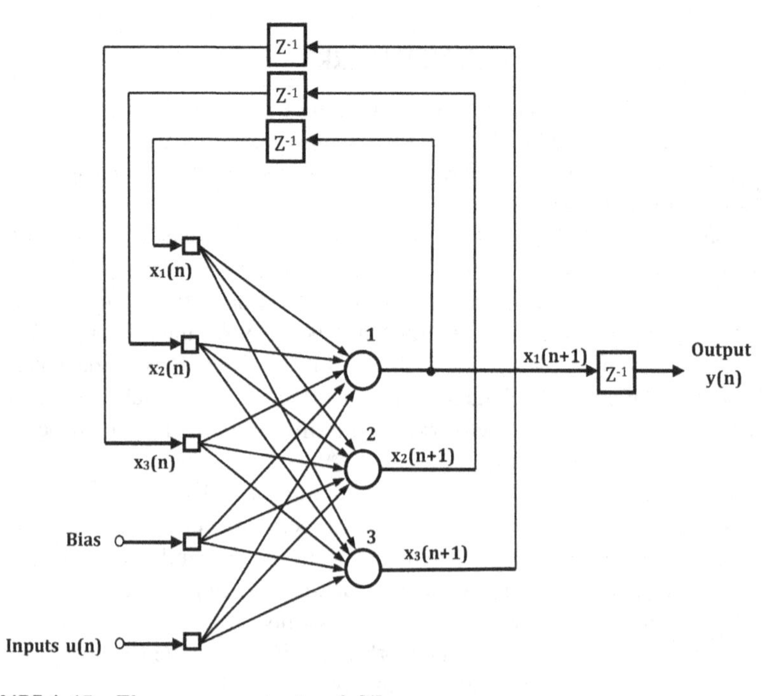

FIGURE A.15 Elman recurrent network [4].

The structure can be extended to "Multi-layer Perceptron with Multi-layer Hidden Layers."

As long as the linearized Elman structure is observable, the original Elman network is considered observable. In this case, the RNN structure can be reconfigured as shown in Figure A.16. In this case, as can be seen, instead of getting feed-back from hidden neurons, outputs (accompanied with their delayed values) are directly fed back to the hidden neurons. This structure is called "Nonlinear Autoregressive Exogenous (or NARX) Network" [4].

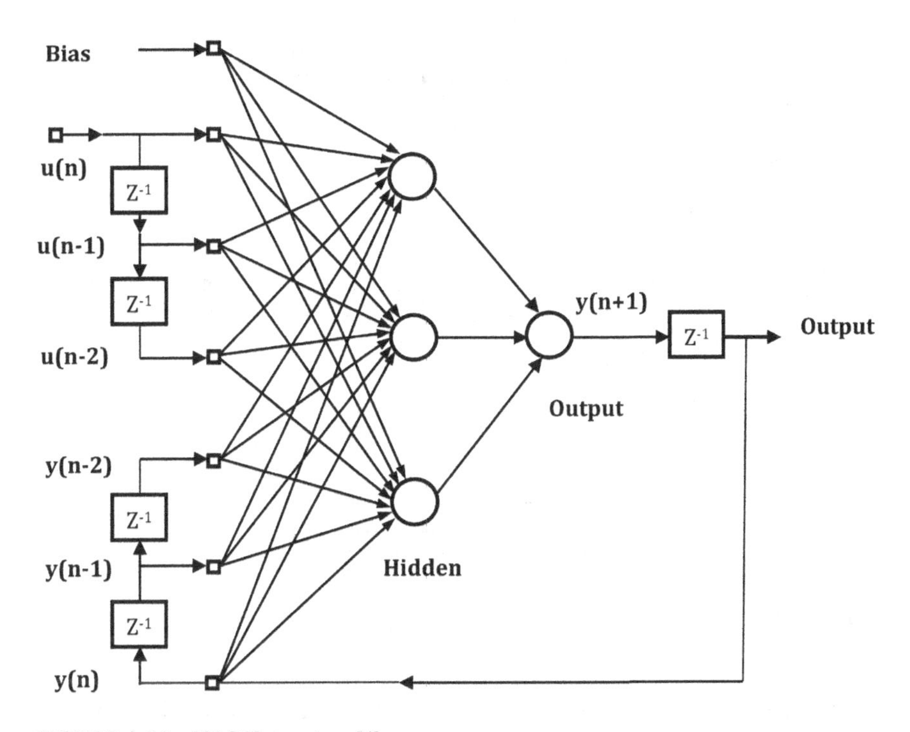

FIGURE A.16 NARX structure [4].

The advantage of NARX structure is that the network training can be much easier than the Elman network as it contains no recurrent links inside the network; if the real system outputs are fed rather than the real model outputs, the network can be trained just as a static neural network. However, the instability issue escalates as to NARX networks. Furthermore, the model stability could not be analytically established and the only way left would be testing.

By extending the Elman structure (which is considered of first order), a more sophisticated network of second order can be obtained. This structure exploits two-by-two multiplication of inputs and feed-back of hidden neurons and is called "Second-Order Recurrent Neural Network."

A.13 DIAGONAL RECURRENT NEURAL NETWORKS

Diagonal recurrent neural networks (DRNN) were proposed in 1995 by Chao-Chee and Kwang, and can be utilized in modeling dynamic systems [12]. In this structure, instead of feeding the delayed outputs of the neurons in hidden layer back to all of them, each neuron's delayed output is merely fed back to itself (Figure A.17).

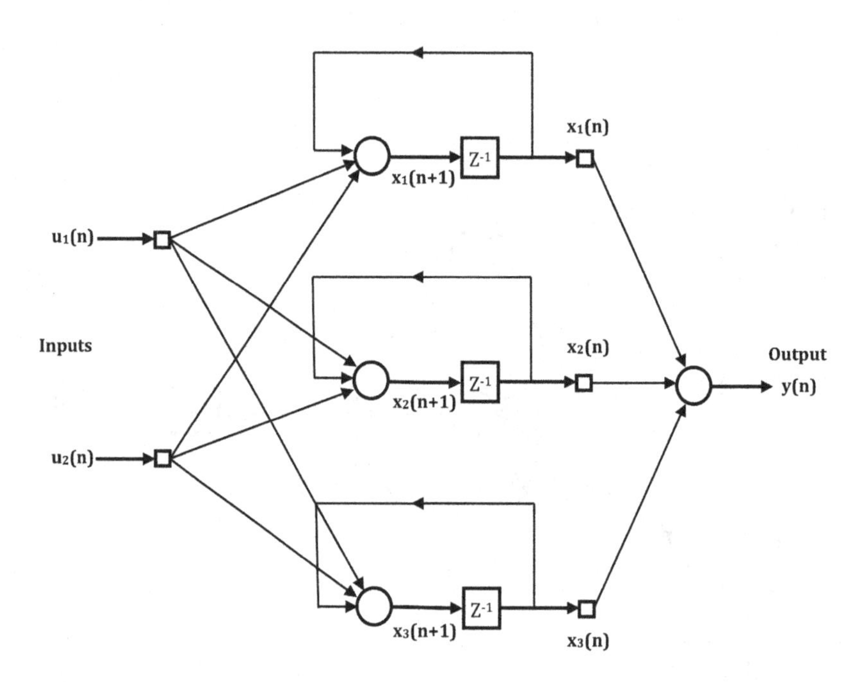

FIGURE A.17 Diagonal Recurrent Neural Network [12].

These networks account for a part of applications, particularly industrial ones. Their merit is a much simpler structure, compared to the globally recurrent network (Figure A.15), and therefore, a much simpler training process. Moreover, unlike globally recurrent networks, they own a necessary and sufficient condition as for stability; feed-back weights have to be of magnitudes smaller than one.

Many systems in industrial applications are approximated via first order time-delayed models. Diagonal recurrent networks are, in fact, an extension to this approach, including a finite number of dynamic sub-models of first order. This means a diagonal state space with totally decoupled dynamics.

Not only is training DRNNs far less taxing, but they also require less hidden neurons compared to RNNs and NARXs. Hence, parameter estimation can be implemented more accurately and easily [12, 13].

This method has been found to be quite effective, particularly, as to systems that can be broken down into inherently first-order subsystems.[iv]

A.14 SUMMARY

In this appendix, methods concerning optimal design of MLP neural networks were investigated. The results obtained can be summarized as follows:

A. Several initial weights need to be tried to ensure convergence to global minimum.
B. The pattern table ought to contain enough data including different operating regions of the system. It needs to be as large and of less noise as possible.
C. Both the networks and the task should be attempted to become as simple as possible.
D. According to bias/variance dilemma, the hidden neurons need to be gradually increased to attain the optimal number.
E. To avoid over-training, appropriate cross-validation set with proper adjustment of early stopping needs to be employed. Network generalization capability is assessed by using a test set. All the sets must be perfect representatives of different behaviors of the system.
F. Depending on the problem in hand, Bayesian regularization, conjugate-gradient method, resilient back-propagation, or Levenberg-Marquardt method can be applied for training.
G. As for each system output, one neural network should be trained.
H. A sensitivity analysis can be utilized in order to avoid applying pointless inputs.
I. To achieve high network performance, the combination of ensemble averaging and improved partitioning can be applied. That is, the system behavior is broken down into different regimes. As for each regime, an extended partition is allocated and as for each partition, a committee of slightly over-fit networks is trained. The connected structure can have a remarkably lower MSE than the optimal single network trained for the whole operating area.
J. With regard to dynamic systems, especially industrial ones, diagonal recurrent neural networks may be more effective than the conventional types like NARX or RNN.

NOTES

i. By error, we mean MSE, or mean squared error, unless otherwise stated. This is calculated as follows:

$$E = \frac{1}{N}\sum_{p=1}^{N} E^p$$

where E^p is the error of every estimated sample:

$$E^p = \frac{1}{2}\sum_{i=1}^{M}\left(d_i^p - y_i^p\right)^2.$$

ii. n stands for the number of weights.
iii. Note that as for stability of a dynamic system, there is not usually a necessary and sufficient condition.
iv. Even as for distributed systems with infinite number of subsystems.

REFERENCES

1. Hagan, M.T., H.B. Demuth, and M. Beale, *Neural Network Design.* 1st ed. 1996, Boston, MA: PWS Publishing Co.
2. Krose, B. and P. van-der-Smagt, *An Introduction to Neural Network.* 8th ed. 1996, Amsterdam, The Netherlands: University of Amsterdam.
3. Nelles, O., *Nonlinear System Identification: From Classical Approaches to Neural Networks and Fuzzy Models.* 1st ed. 2000, New York, NY: Springer.
4. Haykin, S., *Neural Networks: A Comprehensive Foundation.* 2nd ed. 1999, New Jersey, NJ: Prentice Hall.
5. Demuth, H. and M. Beale, *Neural Network Toolbox for Use with MATLAB.* 1998, Natick, MA, USA: MathWorks Inc.
6. He, Y. and C.J. Rutland, *Application of artificial neural networks in engine modeling.* International Journal of Engine Research, 2004. **5**(4): p. 281–296.
7. Brahma, I., Y. He, and C.J. Rutland, *Improvement of neural network accuracy for engine simulations.* Technical Paper 2003-01-3227, SAE International, 2003.
8. Hashem, S., *Optimal linear combinations of neural networks.* Neural Networks, 1997. **10**(4): p. 599–614.
9. Jang, J.-S.R., *ANFIS: adaptive-network-based fuzzy inference system.* IEEE Transactions on Systems, Man, and Cybernetics, 1993. **23**(3): p. 665–685.
10. Nelles, O. and R. Isermann, *Basis function networks for interpolation of local linear models,* in *Proceedings of 35th IEEE Conference on Decision and Control.* 1996. Kobe, Japan: IEEE.
11. Shamekhi, A.-M. and A.H. Shamekhi, *A new approach in improvement of mean value models for spark ignition engines using neural networks.* Expert Systems with Applications, 2015. **42**: p. 5192–5218.
12. Chao-Chee, K. and Y.L. Kwang, *Diagonal recurrent neural networks for dynamic systems control.* IEEE Transactions on Neural Networks, 1995. **6**(1): p. 144–156.
13. Ayeb, M., et al., *SI engine modeling using neural networks,* Technical Paper 1998-02-23, SAE International, 1998.

Appendix B
A Short Review of Some Optimization Algorithms

Optimization simply concerns finding the global minimum (or maximum) of a function over an operating space. It could be constrained or unconstrained, convex or non-convex, single objective or multi-objective.[i]

Equality constraints are called Hard constraints. Yet, inequality constraints are known as soft constraints. Obviously, unconstrained optimization is simpler to solve compared to constrained optimization problems. Consequently, many approaches attempt to convert a constrained optimization problem into an unconstraint problem. For instance, in order to avoid violation of a given constraint, it is converted to a penalizing function (or penalty term) and is added to the objective function. When the constraint is met, the penalty term is (set to) zero, but when the constraint is violated, it takes large values, substantially increasing the objective function value (which is to be minimized). This is called the Penalty Method.

For very simple problems with merely equality constraints, the constraint can be substituted into the objective function. When a continuous differentiable problem contains merely hard constraints, the method of Lagrange multipliers could be employed, to incorporate the constraints into the objective function. In the presence of inequality constraints, the method could be generalized by KKT conditions.

The most famous optimization method as regards the continuous differentiable functions is arguably Cauchy's Gradient-Descent (or Steepest-Descent) method. Gradient-based methods (e.g. gradient descent, interior method, etc.) generally attempt to converge to a local minimum by heading in the opposite direction of the gradient of the function, at each step. Nonetheless, as implied, there is no guarantee for the local minimum found by these algorithms to be the global minimum, unless the function is proved to be *convex*.

When it comes to non-convex (or also non-differentiable) optimization problems, global methods, like some meta-heuristic algorithms, can be employed. However, unlike gradient-based methods (which are accurate approaches), these algorithms are approximate approaches. In other words, gradient-based methods, when analytically converge, yield the exact value of the local minimum in that region. On the other hand, global meta-heuristic algorithms (e.g. population-based or evolutionary algorithms), although attempt to find the global minimum, there is no analytical proof that they converge to the exact value. That is, they are likely to find the *approximate* value of the global minimum.

There are two features describing the quality of a meta-heuristic algorithm: Intensification versus Diversification. Intensification means to keep searching within a candidate region in the operating space. Diversification simply means the ability of an algorithm to search various different regions of the operating space. When finding a better candidate point, Intensification makes the algorithm search the neighborhood thereof more intensely. Diversification, on the other hand, helps the algorithm to leap for other new candidate points (maybe even far from the currently optimal candidate), in order to prevent converging to a local minimum, and thus, to higher the chance of finding the global minimum. Intensification makes the algorithm converge faster, yet diversification causes the convergence to become slower. An ideal algorithm must have both great diversification and intensification capabilities.

Different classifications have been proposed for meta-heurist methods, such as population-based against the single point search, nature-inspired against non-nature inspired, various against single neighborhood structure, memory usage against memory-less methods, etc [1].

Genetic Algorithm (GA) is probably the most renowned and frequently used meta-heuristic algorithm and was first suggested by Holland [2]. GA is a population-based evolutionary method. It consists of at least five steps: initialization, selection, genetic operation, iteration, and termination. In Initialization, a random chromosome population is generated. In selection, using the defined fitness function, the chromosomes with the best fitness are selected as parents. Genetic operations include crossover and mutation, and they intend to generate offspring, and at the same time, perform diversification. In crossover, some parts of chromosomes are exchanged with each other. Yet, in mutation, some parts of chromosomes are randomly changed. This procedure is iterated (from the selection step) over and over until a termination criterion is met.

Particle swarm optimization (PSO) is another population-based algorithm, first proposed by Kennedy and Eberhart [3]. PSO consists of the additive combination of three terms, inspired by the food searching of birds or insects. Imagine a flock of birds coming to a small forest. At first, each would go for a random tree, in search of food (random term). When a tree is devoid of food, the bird goes for another random tree. Meanwhile, the bird's memory helps it not to search repeated trees (memory term). More importantly, when birds (or insects) observe a number of birds (or swarm of insects) gathered on a tree, they conclude that there must be food on that tree, making them fly in that direction (swarm intelligence term). This is in fact a prominent intensification feature of the algorithm, causing it to converge faster than many other methods. Nonetheless, this fast speed, on the other side of the coin, could result in premature convergence to a local minimum.

Differential Evolution was proposed by Storn and Price as an evolutionary algorithm [4]. It consists of at least five steps: initialization, mutation, recombination, selection (and iteration), and termination. It is quite similar to GA, yet the main difference is in the mutation step. The new generation (called the trial vector) in this step is produced using differences between individuals in a simple and fast linear operation called differentiation [5].

Ant colony optimization [6] algorithm, presented by Dorigo, is inspired by the way ants find the best route from colony to food sources. It is especially useful for problems that concern finding the shortest path as a goal [1]. In the real world, when ants explore their environment in search of food, they first wander randomly. After finding a food source, on the way back to the colony, they lay down pheromone trails. When other ants come across such trails, they tend to follow them, instead of random exploration. If they are satisfied with the path, they reinforce the pheromone trail, on the way back, just as well. Interestingly, the pheromone gradually evaporates. Consequently, shorter paths and/or more frequently followed paths will have stronger pheromone trails and gradually dominate.

Artificial bee colony algorithm is another population-based method proposed by Karaboga [7] and is based on the way bees find food sources. There are three types of bees: employed bees, onlookers, and scouts. At first, employed bees search for food around the bee hive, randomly. If one finds a food source, it informs an onlooker and then changes its role to become a scout to search for new sources. Based on the employed-bees' findings (shown in their dance), onlookers decide which food source is better.

NOTE

i. Multi-objective optimization concerns finding Pareto fronts instead of a single solution. Here, only single objective optimization is discussed.

REFERENCES

1. Beheshti, Z. and S.M.H. Shamsuddin, *A review of population-based meta-heuristic algorithm.* International Journal of Advances in Soft Computing and Its Applications, 2013. **5**(1): p. 1–35.
2. Holland, J.H., *Adaptation in Natural and Artificial Systems: An Introductory Analysis with Applications to Biology, Control, and Artificial Intelligence.* 1975, Ann Arbor, MI: Michigan University of Michigan Press.
3. Kennedy, J. and R. Eberhart, *Particle swarm optimization*, in *Proceedings of IEEE International Conference on Neural Networks.* 1995. Perth, WA, Australia.
4. Storn, R. and K. Price, *Differential evolution – A simple and efficient heuristic for global optimization over continuous spaces.* Journal of Global Optimization Volume, 1997. **11**: p. 341–359.
5. Feoktistov, V., *Differential evolution*, in *Differential Evolution: In Search of Solutions.* 2006, Boston, MA, USA: Springer. p. 1–24.
6. Dorigo, M., V. Maniezzo and A. Colorni, *The ant system: Optimization by a colony of cooperating agents.* IEEE Transactions on Systems, Man, and Cybernetics–Part B, 1996. **26**(1): p. 29–41.
7. Karaboga, D., *An idea based on honey bee swarm for numerical optimization*, Technical Report TR06, Erciyes University, Computer Engineering Department, Kayseri, Türkiye. 2005.

Index

A

B

C

D

E

F

G

H

I

J

K

L

M

N

O

P

R

S

T

U

V

W

Printed in the United States
by Baker & Taylor Publisher Services